AUSTRALIA'S DEFENCE

AUSTRALIA'S DEFENCE

DAVID ARCHIBALD

Connor Court Publishing

Connor Court Publishing Pty Ltd

Copyright © David Archibald 2015

ALL RIGHTS RESERVED. This book contains material protected under International and Federal Copyright Laws and Treaties. Any unauthorised reprint or use of this material is prohibited. No part of this book may be reproduced or transmitted in any form or by any means, electronic or mechanical, including photocopying, recording, or by any information storage and retrieval system without express written permission from the publisher.

PO Box 224W
Ballarat VIC 3350
sales@connorcourt.com
www.connorcourt.com

ISBN: 9781925138849 (pbk)

Cover design by Maria Giordano

Printed in Australia

Contents

Foreword	viii
Introduction	x
1. Australian Army	1
2. Royal Australian Navy	25
3. Royal Australian Air Force	57
4. Fuel Security	99
5. China's Coming War	115
6. Going Nuclear	151
7. The Broader Strategic Context	171
8. Funding the Increased Defence Effort	187
Postscript	192
Appendices	
i. The ANZUS Treaty	195
ii. Japan-Australia Joint Declaration on Security Cooperation	199
iii. The case for a new Australian grand strategy	203
iv. The Vast Pacific	207
v. Ambassador Kim Beazley's Address and Remarks, Perth 13th August 2015	212
vi. Step by Step, Here's How to Fight China	222
Notes	244

List of Figures

1. F90 Austeyr Rifle   4
2. Hawkei Light Armoured Patrol Vehicle   11
3. Australian Light Armoured Vehicle (ASLAV)   12
4. Abrams Tank   14
5. Tiger Helicopter   17
6. Bushmaster Protected Mobility Vehicle   24
7. HMAS *Anzac* in Sydney Harbour   26
8. Profiles of the *Soryu, Oyashio, Yuan* and *Collins* Class Submarines   33
9. HMAS *Rankin*   43
10. *Formidable* Class Frigate *RSS Steadfast*   49
11. Cutaway View of the *Canberra* Class Amphibious Assault Ship   51
12. HMAS *Armidale* Patrol Boat   53
13. Sukhoi Su-35   60
14. Saab Gripen   64
15. F-35A   73
16. F/A-18E Super Hornet   80
17. C-17 Globemaster   90
18. E-7A Wedgetail   92
19. JORN Operating Coverage   95

20. JORN Operating Principle	96
21. World Oil Production 1965 – 2030	101
22. Australian Oil Production and Consumption 1965 – 2020	102
23. Australian Refining Capacity 2000 – 2020	104
24. Heating Effect of CO_2 per 20 ppm Increment	110
25. China's Claim Area in the South China Sea	116
26. Chinese Base-building in the Spratly Islands	121
27. Trends in Chinese Vessels in the Waters Surrounding the Senkaku Islands	129
28. Relative Size of Economies in East Asia	137
29. Johnson South Reef, 27th May 2015	147
30. Fiery Cross Reef, 13th July 2015	148
31. Subi Reef, 13th July 2015	149
32. Domestic Grain Production and Grain Imports for the MENA Region	182
33. GDP per Capita relative to US GDP per Capita	183

Acknowledgements

I have had many help me on the way. These good people include Joseph Poprzeczny, John Rice, Chris Mills, Marek Chodakiewcz, who saw potential that no-one else could, Pat who is still serving, and Elizabeth. Thanks to all.

FOREWORD

David Archibald's book, *Australia's Defence*, is an easy, entertaining and informative read. That is a difficult combination to achieve. It describes, in terms anybody can understand, the equipment and personnel of the Australian Defence Forces in relation to what it should do. It does so from a common-sense, layman's point of view. While this may infuriate the professional defense analyst, who may be tempted to pick technical nits with it, it is also the book's greatest strength.

It asks the questions of whether or not China is preparing for war, does Australia have the means to defend itself and what the defence choices are, and makes the answers understandable and accessible. By doing so it opens a debate, which would otherwise be the province of wonks, to the general court of public opinion, where it belongs. Thus it renders an invaluable service, whether you agree with a specific analysis or recommendation or not. For too many people, defence policy is often a matter of faith in institutions, force sizes and strategies that upon closer examination do not deserve it. That is too dangerous to continue.

David Archibald's knowledge of energy issues shines through in the other major subdivision of the book, which is Australia's energy policy. It is hard to avoid accepting his conclusion that Australia's long term energy security decisions, including the hollowing out of its refining capacity, will be nothing short of a noose around its neck in any major international crisis. There is no contingency against disruption. On both sides of the political aisle the preparation for possible disaster consists of a denial that it could ever happen.

But if David's book is an easy but informative read, it is also likely to be a controversial one. It starts from the assumption that China's words and actions should be taken at their face value and implication, which leads to the working hypothesis that war with the Asian giant is brewing on the horizon. Whether true or false, the message "prepare for war" will have a visceral impact on everyone, especially a generation such as ours, which has known nothing but peace. War has become unthinkable, and Archibald wants us to think it: at least in the hypothetical.

To many that will be the intellectual deal-breaker; because war, for obvious and perfectly understandable reasons, is the modern unmentionable; the obscenity that cannot be uttered where the "f" word has long become acceptable. For some it will not matter that Archibald may be right; it is almost as if he had no business being correct and still less business proving it to us. Yet David's arguments must be rejected – or accepted – on their own terms. One cannot, like a wilful child, simply will the nonexistence of unpleasant things. The world is, what it is. It pays to remember Leon Trotsky's famous dictum: "you may not be interested in war, but war is interested in you."

Therefore, whether one likes it or not; whether it confirms every suspicion or rouses the reader to indignation, David Archibald's book is worth a read. During the height of the Cold War, movies like *Dr. Strangelove* and books like *Fail Safe* terrified the reader with fiction. Surely we can bear being terrified by factual argument, if only because like the "Dr. Strangelove" of history, our greatest safety lies in being able to think the unthinkable, the better to avoid it.

Richard Fernandez, The Belmont Club

INTRODUCTION

Sometimes it is best to take people at face value. If they say they are going to do something, it would be prudent to plan and to act on the basis that they may very well end up doing what they say they will be doing, even though a dispassionate analysis would suggest that what they will be doing will be imprudent to the point of being stupid and self-destructive. These imprudent people might be quite intelligent, industrious, lack for nothing materially and yet none of that would hold them back from their heart's desire, which is to have respect born of fear. If these imprudent people are actually physically preparing to do what they say they are going to do, that is all the more reason to believe that they will be doing what they say they are going to do. If these physical preparations mean they are diverting resources from solving other problems that might seem pressing, such as having clean air and clean water, then they really want to do what they say they are going to do.

Which brings us to the subject of China. Following the collapse of the Soviet Union and the end of that regime's policy of having ideological hegemony over all the earth, the world scene was like an illustration from a creationist textbook, with all the animals of the earth living in harmony and not one jealous of what any other might have. But, as Samuel Huntington predicted in *The Clash of Civilisations and the Remaking of the World Order*, the end of the age of ideological conflict mean that civilizational conflict, the normal state of affairs in the world, would reassert itself.[1] And so it has. The Middle East, the source of one third of the motive power of the world, now has four civil wars running concurrently. There are many deeply unhappy people in the Middle East, not content in their own spiritual fulfilment, who want to force their beliefs on others. But what is required to keep the unhappiness of the Middle East from affecting the civilised world is a mere police action by comparison the effort needed to thwart China's sense of grievance.

We should be taking the Chinese at their word. When they say they will seize the Ryuku island chain from Japan, then they will be attacking Japan. When they say that they are going to seize the South China Sea, then they will be killing Vietnamese, Filipinos, Indonesians and Malaysians to do so. When China says that they will be having a war with the United States in a Götterdämmerung that will decide who rules the world thereafter, then another world war is coming. They have said all those things. Their physical preparations for all these wars are all the more reason to take them at their word.

Australia's experience of World War I was traumatic – our combat deaths were more than those of the United States. Fighting for national survival during World War II was also traumatic and another generation gave their lives that we might live freely today. Both those wars were preceded by the drums of war pounding out their warning. At least they were heeded in the latter case and Australia started preparing a few years in advance. Once again the drums of war are pounding. The longer we neglect to heed the message from the war drums, the greater our suffering will be in the coming conflict. As recent anniversary celebrations have reminded us, it is now 70 years since the end of World War II and civilisational memories of the death and destruction, privation and suffering of that war continue to fade. The institution of our defence forces has shrunk to being a sort of holdover from a previous age, for some reason necessary to have in a modern state but not so necessary that it should be organised and funded to the level of actually being effective. The defence of the realm is the first responsibility of the state. Running a country is like good household budgeting – the priorities go to the top of the list and the indulgences fall off the bottom. There will always be enough money for defence, it is just that we don't have our needs properly prioritised yet.

This book is all about China. Apart from Australia's Children's Crusade-type effort in fighting Islamic State, there would be otherwise no need to awake from our torpor. China's coming war will upend things for all of us. Will Chinese submarines be able to sink most of the oil tankers coming to Australia so that everything grinds to a stop? Even to the point that the food distribution system breaks down? What will we do with all the Chinese internees for the duration? We currently have a major security burden in stopping Islamic terrorists from poisoning our dams and similar atrocities. Will all the Chinese agents in Australia be doing that on the outbreak of hostilities, and other acts of mayhem and destruction?

There are many necessary and practical things we should be doing in preparation for what is coming. We have the wrong sort of tanks and certainly not enough of them. Successive governments, Labor and Liberal, have shied away from buying the right sort of battlefield transport for our diggers. They just don't want to pay up, and young Australians will die unnecessarily. We don't have enough troops in the first place. Our decrepit submarines need replacing as soon as possible, and making the right choice may be existential. If there is one choice that is existential, it is our choice of fighter aircraft. The F-35s we are heading towards promise to reprise the role of our Boomerang fighters in the defence of Darwin in 1943. They will simply be shot out of the sky on their first mission. We have only contracted to buy two F-35s though. We can still back out of buying any more than that and instead get another aircraft that is fit for the purpose of defending Australian airspace. The chapter on the RAAF explains all.

Fighting the next war with China will be a whole-of-nation exercise. This book is a primer on what to expect.

David Archibald

1

AUSTRALIAN ARMY

History

The first humans entered Australia about 40,000 years ago, which was 10,000 years after they left Africa. Proto-humans had occupied Asia close to our north for hundreds of thousands of years prior to that but were displaced by the modern humans coming through. One exception was the hobbit people of the Island of Flores in Indonesia who lasted up until at least 13,000 years ago. It is thought that there were many pulses of Aboriginal peoples arriving over the last 40,000 years in Australia, reflecting the current physical diversity of Aboriginal tribes. Based on genetic evidence, one of the last pre-European arrivals in Australia was from southern India about 4,230 years ago[1]. This date also coincides with many changes in the archaeological record of Australia, which include a sudden change in plant processing and stone tool technologies, with microliths appearing for the first time, and the first appearance of the dingo in the fossil record. The group from India contributed 11 per cent to the gene pool of Aborigines of northwest Australia at least.

British settlement of Australia began in 1788 with protection provided by British Army garrisons up until 1870 when the garrisons were withdrawn. By that time the colonies each had their own militia units. The first foreign war in which Australian-born soldiers served was the Maori Wars of 1863 to 1872, in New Zealand colonial units or the British Army. The colonies employed professional soldiers in coastal artillery units and by the early 1880s were holding inter-colonial defence conferences. The first official foreign expedition by an Australian military unit was an infantry battalion sent from New South Wales in 1885 to the British campaign in the Sudan.

The first Australian army unit was created on 24th August 1899 by the merger of the colonial artillery units in preparation for Federation on 1ˢᵗ January 1901. Three months later, the rest of the colonial forces were brought into the new Australian Army structure. In the meantime, Australian forces served in the Boer War and the Boxer Rebellion in China. Then came World War I in which 331,814 personnel served overseas as part of the Australian Imperial Force (AIF). This was 13 per cent of the Australian male population. Of these, 18 per cent (61,859) were killed or died with a total casualty rate of 64 per cent. This was from a population of less than five million. World War II was less tragic with 18,000 dead out of almost 400,000 who served overseas. Australia has been at war on and off thereafter with the Korean War, Malayan Emergency, Vietnam War, Gulf War, East Timor, Afghanistan and Iraq War. At the time of this writing, Australian air and ground forces are back in Iraq as part of a desultory campaign against Islamic State.

Structure

The main force of the Australian Army, of about 29,000 regular soldiers, is currently being restructured into three combat brigades under a plan called Plan Beersheba. This will take place over the three years to 2017. The brigades will incorporate organic artillery and armour. The most basic infantry unit is the section. Historically the section consisted of two scouts, the section leader with the rank of corporal, a two-man machine gun team and five riflemen for a total of ten men. Under the Army's Plan Beersheeba, the section will be eight men further sub-divided into two groups each colloquially known as a 'brick' – a four man team consisting of a marksman, a machine gunner, a rifleman with a grenade launcher attachment and either the section commander or the second-in-command. From there it goes up in multiples of three such that three sections make a platoon with the addition of a lieutenant, sergeant, medic and platoon radioman. Three platoons make a company which is commanded by a major, a captain and a warrant officer. A company-size body of men, up to 120, is the limit that men will recognise as being part of the same group. This is important in developing trust and confidence in battle. The next step up is the battalion consisting of four rifle companies, a heavier supporting arms company (mortars, anti-tank weapons and heavier machine guns) and an administration company for a total of about 700 men. Three battalions make a brigade which, with its supporting elements, would be about 2,500 men. Three brigades make a division and three divisions make a corps. As well as those units, there is the regiment which can range in size and purpose in the middle of that structure. For example there are artillery regiments, combat signals regiments and other specialisations.

So Australia, a nation of 24 million, will have three brigades of full time troops totalling about 7,500 men available for land warfare. These brigades will be developed from 1st Brigade in Darwin, 3rd Brigade in Townsville and 7th Brigade in Brisbane with each having the following standardised components: a brigade headquarters, an armoured cavalry regiment, two standard infantry battalions, an artillery regiment, a combat engineer regiment, a combat service support battalion and a combat signals regiment. Apart from the three combat brigades, there is also 6th Brigade (Combat Support, Intelligence, Surveillance, Target Acquisition and Reconnaissance), 16th Aviation Brigade and 17th Combat Service Support Brigade.

Australia's special forces are in a separate structure called Special Operations Command (SOCOMD) which includes the Special Air Service Regiment, two Commando Regiments and their supporting elements.

Australia's deployable professional field force is pathetically small. The philosophy behind the deployment of the three combat brigades is that one will be actually deployed while another is undergoing refit after deployment while the third is being trained for deployment. So the plan is that a nation of 24 million will have only one ten thousandth of that number available on the front line at any one time.

So how big should it be? First of all, bear in mind that the army has been allowed to shrink to its current size without any analysis of what is required to defend the continent. The western half of the continent doesn't have a deployable regular army unit above company size, for example.

Australia needs at least three times the number of combat brigades that it intends to have. The real requirement may be much higher. In any extended conflict between nation states in our region, our current tiny force will be chewed up quickly. It is also at great risk by virtue of being so tiny and thus without backup.

Weapons

The basic infantry weapon in the Australian Army is the F88 Austeyr which was developed from the Austrian Steyr AUG STG-77 assault rifle. Introduced in the late 1980s, the rifle uses an Australian version of the 5.56 mm NATO round. The NATO round shrank from 7.62 mm (0.308 inches) to 5.56 mm (0.22 inches) in 1980 as it was thought, based on analysis of firefights, that most fighting took place within 300 metres. The smaller round becomes less effective after 200 metres but that is offset by the much greater number of rounds that can be carried. For standard NATO rounds, the 5.56 mm cartridge weighs 12.3 grams while the 7.62 mm cartridge is just over twice that at 25.4 grams, with the rounds weighing 4.02 grams and 9.33 grams respectively. The 5.56 mm round has a muzzle velocity of 970 metres per second while that of the 7.62 mm round is 838 metres per second. Kinetic energy is mass

Figure 1: F90 Austeyr Rifle
The Australian Army has ordered 30,000 of the recently upgraded version of the Austeyr rifle, shown here with the SL40 grenade launcher attached. The magazine holds thirty 5.56 mm rounds. Muzzle velocity is 970 metres per second and the effective range of the 4.8 kg rifle is 300 metres.

by the square of the velocity. So while being under half the weight of the 7.62 mm round, the 5.56 mm round has 60 per cent of the kinetic energy. That said, modern engagements tend to be fought at greater distances which favours the larger calibre. Some of the fanatical Taliban in Afghanistan kept fighting after being hit with 5.56 mm rounds and returned fire with 7.62 mm rounds.

Australian troops use full metal jacket ammunition in which a soft core of lead is encased in a shell made of harder metal. Signatories to the Hague Convention of 1899 are prohibited from using hollow point bullets that are designed to expand or flatten upon hitting soft tissue. The United States did not sign the section of the convention related to bullet types and the U.S. Marine Corps has switched to hollow point ammunition after experience in Afghanistan. For example, a full metal jacket 9 mm round has a one-shot stop rate of 70 per cent while a 9 mm hollow point has a one-shot stop rate of 91 per cent. Adopting hollow point rounds, also referred to as open-tip rounds, compensates for some of the ineffectiveness of 5.56 mm rounds.

The Austeyr is a bullpup design which means that the magazine is behind the trigger instead of in front of it as in most assault rifles. This makes the rifle shorter and in theory this should make it more useful in fighting at close quarters. The most recent Israeli rifle is the Tavor which is also a bullpup design in 5.56 mm. The Israelis developed the Tavor because of their close-quarters fighting in urban environments and carrying it in mechanised infantry vehicles. The bullpup design allowed the weapon to be compact while keeping a longer barrel able to achieve high muzzle velocities. Under a $72.9 million contract recently signed with Thales Australia (owned by the French aerospace and armaments company Thales), the Army will be getting 30,000 copies of an enhanced version of the Austeyr and 2,500 SL40 grenade launchers. The new rifles will come in two barrel lengths, a 16 inch carbine and a 20 inch rifle, and cost about $3,000 each. They will be made in the former small arms factory in Lithgow, now operated by Thales Australia.

The next weapon up is the FN Minimi which is the light machine gun developed in the mid-1970s by Fabrique Nationale of Belgium. The

Australian Army version is made by Thales Australia, and designated the F89 Minimi. It fires 5.56 mm ammunition. The Australian Army also has a version in 7.62 mm ammunition called the Maximi. The largest machine gun in the Australian Army is the Browning M2 .50 calibre (12.7 mm) with an effective range of over 2,000 metres. The original design dates from the end of World War 1, making it almost 100 years old.

The mortar the Australian Army uses is the F2 81 mm Mortar, which is the Australian version of the British L16 81 mm mortar. This was designed in 1956 and has been in service from 1965. The effective range is from 100 metres out to 5.6 kilometres. The 81 mm mortar round weighs 4.2 kg. If the mortar is to be man-packed into battle, it is carried disassembled into three loads of the barrel, baseplate and bipod with sights. Each load is approximately 11 kg, not including mortar rounds. Other soldiers supporting the mortar squad would carry up to four rounds each.

These days combatants are detecting each other at greater ranges and generally the side that can engage and disengage at will has the advantage. That means having a longer range mortar. The next size up in mortars is the 120 mm with a range of 8.2 kilometres. Range can be extended to 17 kilometres with rocket-assisted projectiles. These give the 120 mm mortar the range of medium artillery at a much lower weight to be carried into battle. There are now GPS-guided 120 mm mortars and these can effectively replace medium-range artillery if rocket-assisted.

On patrol, an infantryman will carry 120 rounds for his rifle. In a section, the machine gunner and his number two will carry 200 rounds of belted or drum ammunition each. Everybody carries two fragmentation grenades and each section will have six smoke grenades.

Artillery

The main artillery piece of the Australian Army is now the M777 howitzer. Introduced in 2010, it is a 155 mm towed howitzer which sells for US$3.7 million. The M777 is manufactured by BAE Systems at its facility in Hattiesburg, Mississippi. About 70 per cent of the howitzer is from US-made parts with the balance from the UK. It replaced the M198 howitzer in the Australian Army. At 4.1 tonnes, it is 42 per cent lighter with most of the weight reduction due to the use of titanium. The minimum gun crew for the M777 is five compared to the M198's nine personnel. Australia has only 54 M777s, making up six batteries of eight howitzers. So each multi-role armoured brigade will have the fire support of 16 howitzers. That total inventory of howitzers amounts to just over one per half million Australian citizens. It is completely inadequate.

The effective firing range of the M777 is 24 kilometres. Using base-bleed rounds for extended range increases that to 30 kilometres. Base-bleed increases the range of artillery shells by about 30 per cent. While most of the drag on an artillery shell comes from its nose pushing air aside at supersonic speeds, another source of drag is the vacuum left behind the shell due to its flat base. Base-bleed overcomes this drag by using a small gas generator in the base of the shell. Not much thrust is produced but the drag due to the vacuum is dramatically reduced by filling the area behind the shell with air pressure. There is a slight decrease in accuracy due to the more turbulent airflow and small reduction in explosive payload due to the space taken up by the gas generator. Base-bleed technology was originally developed for Swedish coastal artillery in the 1960s.

The Army has stocked the M982 Excalibur round and XM1156 Precision Guidance Kits for the M777s. The Excalibur is an extended range shell with GPS guidance. The extended range of up to 57 kilometres is achieved by the use of folding glide fins which allow it to glide from the top of its ballistic arc towards the target. In February 2012, a US Marine Corps M777 howitzer fired an Excalibur round which killed a group of insurgents at a range of 36 kilometres. Range testing of Excalibur shells has shown that they hit within an average of 1.6 metres from the target. Due to that level of accuracy, one Excalibur round is equivalent to the use of between 10 to 40 unguided artillery rounds in eliminating a target. In April 2008, the Australian Army ordered 680 Excalibur rounds for US$58 million at an average cost of US$85,000.

In 2013, the Australian Army also ordered 4,002 M1156 Precision Guidance Kits for US$54 million. This translates to US$13,500 per kit but includes training and associated equipment. The value of these kits comes from the fact that an ordinary 155 mm artillery shell has a circular error probability of 267 metres at its maximum range of 24 kilometres. This means that a shell could land up to 267 metres from the coordinates that it was aimed at, making it dangerous to call for close artillery support at long ranges. The Precision Guidance Kit (PGK) screws into the nose of the artillery shell in replacement of the conventional fuze. It has GPS guidance and vanes to control the flight of the shell. Within five seconds of being fired, the PGK checks to see whether or not it will land within 150 metres of the aim point. If it thinks it isn't going to, it won't explode. The purpose of this feature is to give troops more confidence in calling in artillery support close to their position.

In comparison to Australia's total of 54 M777 howitzers in service, Singapore has 225 howitzers of that calibre plus a further 50 self-propelled 155 mm howitzers. With a quarter of Australia's population, Singapore has more than five times as many 155 mm howitzers. Singapore has 20 times the artillery intensity of Australia on a per capita basis. Why would that be? The reason is that Singapore understands what is needed on the battlefield to survive and Australia has made only a token effort.

Take the example of the self-propelled howitzers. Singapore decided to develop their own self-propelled howitzers because most Western howitzers were too wide for the narrow bridges of Southeast Asia. So their version is only 3 metres wide. The important thing is that they have some. Australia is the only modern nation without any self-propelled howitzers. We were going get a few under an acquisition programme that started in 2006. The choice came down to the German PzH 2000 and the South Korean K9 Thunder with the latter being the only one that satisfied the Army's requirements.

The evaluation process for the self-propelled howitzer reflects poorly on the Department of Defence and the Minister at the time, Stephen Smith. A decade ago, Australia set out to acquire a token force of up to 24 self-propelled howitzers. Tenders initially closed in April 2008 but no decision was made. Following that, the South Koreans were encouraged to conduct field trials in Australia at their expense. No contract followed from that either. The South Koreans were treated very casually. And Australian troops will die as a consequence of the decision not to acquire self-propelled howitzers.

When artillery is facing possible counter-fire by an enemy artillery battery, they usually can only fire a few rounds before the enemy artillery is able to calculate where they are firing from and subject them to counter-fire. This counter-fire has little chance of scoring a hit on an artillery piece that would disable it. So what is fired are cluster rounds in the hope of killing or disabling the crew. The self-propelled howitzers are armoured so that the crews can survive the counter-fire. Australian artillery won't have that armour so it will suffer a higher death rate. The Army has acknowledged this. Australia signed up to the convention banning the use of cluster munitions. Our most likely antagonist, China, is not a signatory to that convention. So it could come to pass that Australian artillery might be firing at Chinese artillery and not having much effect on them while the Chinese are firing back with cluster rounds and killing our artillery crews.

How much artillery should Australia have? If Singapore, with less than a quarter of Australia's population, has a total of 275 howitzers sized at 155 mm, does that mean that Australia should have at least 1,000 howitzers at that calibre? Yes, spread over the number of combat brigades that Australia should have, that is probably near to the right figure. As to the ratio between self-propelled and towed artillery, one guide is what the Israeli army has which is 650 to 300, a ratio of about two to one.

Australia should also adopt GPS-guided rockets. To put the evolution of precision rockets into context, smart bombs dropped from aircraft were first used in the 1991 Gulf War. They were more effective than artillery and that led to a shift away from using artillery. In 2004, a GPS-guided version of the Multiple Launch Rocket System (MLRS) was introduced as GMLRS. Like the unguided version, the GMLRS are packaged and used in containers (pods) holding six rockets each. The GMLRS rockets cost about US$110,000 each and have been very successful in Afghanistan. It has a maximum range of 85 kilometres and lands within metres of its target. The GPS guidance has an inertial backup system.

The 85 kilometre range means that one GMLRS system can support a front 170 kilometres wide. By comparison, the Excalibur (GPS-guided 155 mm shell) has a maximum range of 37 kilometres and 120 mm mortars have a range of about 7.5 kilometres. A pod of six GMLRS can be carried on an 11 tonne truck. This is the HIMARS (High Mobility Artillery Rocket System) which can be carried in a C-130 aircraft.

The GMLRS missile has a diameter of 227 mm and weighs 309 kg. The warhead weighs 89 kg, about half of which is explosive. A 155 mm artillery shell has 6.6 kg of explosives. The 120 mm mortar shell 2.2 kg of explosives. So in all, the GMLRS system mounted on a truck has over five times the footprint of Excalibur-equipped 155 mm artillery and delivers seven times the amount of explosive at approximately twice the cost - all with the same accuracy. The mere presence of GMLRS on the battlefield will reduce enemy ability to concentrate forces well behind the front lines. It is important to have an artillery system that outranges the enemy's artillery and the GMLRS does that.

Each HIMARS truck costs US$3.5 million of which US$3 million is for the launcher and $0.5 million is for the truck. Adding the missiles takes the cost up to US$4.2 million. Of course Singapore, which is much more intelligent about its defence acquisitions than Australia, acquired the HIMARS system in 2007, not long after it proved so useful in combat. Singapore has 18 of them. How many should Australia obtain? Probably at least five times as many as Singapore. Depending upon the supporting elements and the number of reloads, the cost would be in the range of $1 to $2 billion. It would make the Australian Army far more effective in the field and well worth the cost.

And this is how it could work in battle. Australian artillery could open up and attract counter-battery fire. Once the location of where the counter-battery fire is coming from, HIMARS batteries located to the rear and out of range of enemy artillery will reply with cluster munitions. We could eliminate enemy artillery without too much danger to Australian crews. At the moment the situation, with our exposed crews, is the reverse of that.

Protected mobility

There are three levels of armour on the battlefield, starting with protected mobility. In Australian service, that is the Bushmaster vehicle and the M113 armoured personnel carrier (APC). The Bushmaster is a four-wheeled vehicle weighing 12.4 tonnes and costing $0.58 million. It has a driver and up to nine passengers. A remote weapon station can be fitted on the roof. The Bushmaster is manufactured in Bendigo by Thales Australia. The Army has over nine hundred in service. Thales' Bendigo factory is also waiting on a contract to produce 1,300 of the Hawkei light armoured patrol vehicle to replace the Army's Landrovers. The Hawkei weighs 7.5 tonnes and has a fuel economy of three kilometres to the litre.

The M113 is a tracked vehicle weighing 12.3 tonnes with a crew of two and can take up to 10 passengers. It has been in service with the

Figure 2: Hawkei Light Armoured Patrol Vehicle
The Hawkei is a successful Australian design. To be built in Bendigo, up to 1,400 could be supplied to the Australian Army. A ring-mounted machine gun or remote weapons station can be installed in the roof.

Australian Army since 1965. Of the 700 M113s in Australia's inventory, 430 were upgraded at a cost of about $600 million.

The Bushmaster can survive rocket propelled grenades and heavy machine gun fire. Vehicles as large as the Bushmaster though can be detected at long ranges, say four kilometres, and engaged at that sort of range. Recent experience in Syria shows how effective US antitank guided missiles, in this case the BGM-71 TOW, is against armoured vehicles moving at high speed.

Infantry fighting vehicle

The next level up is called the infantry fighting vehicle (IFV). The role of the IFV is to carry infantry into the battle zone and also to provide fire support with a cannon of at least 20 mm calibre. IFVs can take on main battle tanks if they are also equipped with anti-tank missiles. By comparison, protected mobility vehicles such as the Bushmaster and the M113 are not expected to engage enemy armoured vehicles.

In the Australian Army, the vehicle currently filling the IFV role is

Figure 3: Australian Light Armoured Vehicle (ASLAV)
The main armament of this infantry fighting vehicle is a 25 mm chain gun mounted in an electrically operated turret. It also has two 7.62 mm machine guns.

the Australian Light Armoured Vehicle (ASLAV) with 257 in stock. This is an eight wheeled, amphibious vehicle weighing 13.5 tonnes. It has a crew of three and can carry up to six passengers. Main armament is a 25 mm chain gun, with two 7.62 mm machine guns and a 76 mm grenade launcher. The Army has started an evaluation process to replace the ASLAV. Its replacement is likely to weigh at least twice as much. More armour is required on the modern battlefield to survive the proliferation of 30 mm cannon on enemy IFVs. The replacement is likely to be tracked so that it can keep up with tanks.

Australia has been delaying the purchase of a new IFV because of two factors. The first is actually having to spend the money and successive governments would rather just kick the can down the road. The second is a sort of conditioned helplessness. The Army just keeps studying the problem instead of choosing. There are a number of good IFVs available that would be quite suitable, but we are unlikely to do better than just plumping for the Swedish-made CV90. This is a 28 tonne tracked vehicle that has a crew of three and carries seven infantry. With a top road speed of 70 kilometers an hour, the CV90 can go 300 kilometres on internal fuel. The vehicle turret carries a 30mm autocannon and a coaxial 7.62mm machine-gun. Also in the turret is a thermal imager for night operations. The vehicle armour protects against projectiles of up to 30 mm calibre.

The CV90 is 6.55 meters (20.3) feet long and 3.1 meters (9.6 feet) wide. Average cost of a new CV90 is about $5 million each. The Swedish army has 509 of them, Switzerland 185, Netherlands 193. All these countries have much smaller populations than Australia. How many should Australia get? An appropriate starting number would be the number of M113s that we acquired back in the 1960s – about 700. All up it would be about a $5 billion exercise.

Tanks

The next level up in armoured mobility is the main battle tank. Tanks are very useful on the battlefield. Making an advance without tanks is three times as expensive in infantry as making an advance with them. One

Figure 4: Abrams Tank
The Australian Army has 59 of these tanks. Our nearest neighbour, Indonesia, has 103 Leopard 2 tanks. The Leopard 2 is equally as effective in combat but with half the fuel consumption of the Abrams.

of the finest examples of the utility of tanks on the modern battlefield is the Battle of 73 Easting. This was a tank battle on the 26th February, 1991 during the first Gulf War. A scouting element of 27 M1A1 Abrams main battle tanks with 36 Bradley infantry fighting vehicles encountered a dug-in Iraqi armoured position. Iraqi losses were 85 tanks, 40 tracked vehicles, 30 wheeled vehicles and two artillery batteries. The US loss was one Bradley. The overwhelming victory is ascribed to surprise, training and a qualitative edge in the US equipment. A similar result can be expected any time there is a similar matchup. That is why it is important to have the training and qualitative edge in tank forces – it is decisive on the battlefield.

Australia bought 59 reconditioned M1A1 Abrams tanks in 2007 at a cost of $560 million to replace 101 Leopard 1 tanks that had been bought in 1974. The Leopard 1 tanks had replaced 143 Centurion tanks. This is a common theme in Australian defence procurement of reducing the number of a type of weapon while at the same time buying the wrong replacement. The Leopard 1 tanks should have been replaced by twice the number of Leopard 2s, not nearly half the number of Abrams.

The Leopard and Abrams tanks are twins that were separated in utero. In 1974, the United States and Germany signed a memorandum of understanding on the possible joint production of a new main battle tank. The US Army tested its Abrams prototype against the Leopard 2 prototype in 1976. The main difference between the two is that the Abrams has a gas turbine engine whereas the Leopard 2 has a diesel engine, with twice the fuel economy of the gas turbine. The Leopard 2's fuel capacity of 1,160 litres gives it a maximum road range of about 500 kilometres. The Abrams' 1,900 litres of fuel will take it 426 kilometres.

One litre of fuel will take a Leopard 2 431 metres and an Abrams only 224 metres – just over half the distance. Starting the Abrams takes 38 litres of fuel. It also consumes 38 litres per hour when idling. You might think you could live with all that but the Abram's high fuel consumption means that it has twice the logistic tail of the Leopard 2. The high fuel consumption is also operationally restrictive. The hot exhaust of the gas turbine engine also means that troops can't follow on foot behind the Abrams and benefit from the protection provided by its bulk.

In short, the Abrams is a dog. Thankfully we have only 59 of them. It won't be such a great loss to park them up. They don't have much value in the second hand tank market – the United States has 4,000 in storage in the Nevada desert. Intelligent tank buyers get the Leopard 2. Our near neighbours to the north, Singapore and Indonesia, have 212 and 103 Leopard 2 tanks respectively. A number of Leopard 2 buyers have made some great bargains. For instance Chile bought its Leopard 2A4s for just 250,000 Euros each. New build Leopard 2s to the latest standard, the Type 2A7, cost about 10 million Euros. The troubles in Eastern Europe have tightened the tank market. Germany is now making new-build Leopard 2s for itself instead of selling off stocks. The Leopard 2A7 weighs close to 70 tonnes. As recently as 2006, the Leopard 2A6 weighed 55 tonnes. Increased lethality on the battlefield has required more armour, especially on top.

That only leaves the question of how many we should be acquiring. If Singapore, with just under a quarter of our population, has 212, does that mean that we should be getting at least 800? Australia's land mass of

7.6 million square kilometres makes it the sixth-largest in the world. If we had 800 tanks, that would amount to approximately one per 10,000 square kilometres – a block 100 kilometres square. It may be about the right number. If we had two armoured brigades, that would take about one third of the number and the rest would be spread over the combat brigades, reserve units and tanks held in stock. The cost would be equivalent to that of Australia's three new Hobart-class destroyers – or one third of the annual cost of funding Aboriginal welfare. The funds exist for the purchase of the tanks we need, as they do for all the other things we need. It is just a question of getting our priorities right. Attacking forces have to be at least three times larger than defending forces to prevail, so the larger Australia's land army is, there is a geometric progression in the size of the invading fleet needed to defeat it.

Israel has one-third of Australia's population and 4,170 tanks. Countries that have recent experience of fighting wars and want to survive have lots of tanks – they haven't gone out of fashion. It is also instructive to see how the Israelis view artillery.

Helicopters

Helicopters have three roles in the Army: transport, reconnaissance and attack. The ARH Tiger now fills the reconnaissance and attack roles. It has suffered low availability due to a poor parts stocking policy by the Army. It seems that the Army expects the supplier to keep parts in stock whereas the maintenance contractor, Australian Aerospace which is part of Airbus, relies upon getting parts from France as needed. The Tiger performed well in service with the French contingent in Afghanistan. Australia has 22 of them.

In transport helicopters, Australia has an inappropriate and expensive helicopter fleet due to one of the bad decisions of the Howard government. It started as a reaction to the logistics problems of the East Timor deployment in 1999 which revealed a need for additional troop lift helicopters that were capable of operating off Navy ships. The government thought that it would be better to have commonality

Figure 5: Tiger Helicopter
This is the European version of the Apache helicopter. Both these types were originally created to fight Soviet tanks on a European battlefield. Australia has fewer than one per million head of population.

between the Army helicopter fleet and that of the Navy. The choice was between the NH90 from Airbus in Europe or the latest version of the Black Hawk helicopter from Sikorsky in the United States, the UH-60M. The Black Hawk was already operated by the Army and so our defence forces were already familiar with it. The Australian Army, Navy, Air Force and the Chief of Defence all advised the government to buy the Black Hawk, which was also US$600 million cheaper. Instead, the government signed up for the Airbus offering of 46 helicopters at a cost of $4.013 billion. That works out to $87 million a copy.

The NH90 was a new design which is designated the MRH90 in Australian service. It entered full-rate production before the completion of flight trials and type certification. In their sales blurb, Airbus claimed that the NH90 would need only 2.6 maintenance man-hours per flight hour. Maintenance in actual practice is an order of magnitude higher, down from a peak of 97 man-hours in January 2012 to 27 in April 2014. Airbus built the NH90 with 20 seats which turned out to be too narrow to take fully-equipped troops. The number of seats has been reduced to 14, including the two aircrewman seats. As at May 2013, the MRH90 Program Office in the Australian Department of Defence had identified

35 requirements that the MRH90 design and construction had failed to achieve during the period April 2013 to April 2014. For that period, the average percentage of aircraft serviceable in the MRH90 fleet was 47.6 per cent. The worldwide serviceability of the NH90 was 38 per cent, averaged over the year to November 2013, so it seems that the maintenance problems of this helicopter are inherent to its design. The Department of Defence made a claim of liquidated damages against Australian Aerospace, the local offshoot of Airbus, over the NH90's shortcomings. As part of the settlement, Airbus gave the Department of Defence another helicopter to take the total to 47. But the joke was on the Department of Defence because of what Airbus charges for spare parts. Airbus just increased its profitability by giving the Department of Defence another helicopter to be serviced.

To quote an Australian National Audit Office (ANAO) report on the MRH90, "By May 2011, DMO found the MRH90 spares to be significantly more expensive than equivalent spares purchased for the Black Hawk helicopters through US Government Foreign Military Sales (FMS). An extreme example is an MRH90 plastic plug, which costs $2.18 through FMS, and cost $753.30 when acquired from Australian Aerospace. Similar price mark-ups occurred for the ARH Tiger aircraft, for which a wheel locking pin cost €5,783.63 when a similar pin for Black Hawk aircraft cost A$9.67."[2] The ANAO requested Defence advice on the result of audits or cost investigations carried out to assess the extent of such price differentials. Defence informed the ANAO in April 2014 that 'no specific audits or cost investigations that include the cost of role equipment had been undertaken". Can you imagine the mirth in the Airbus head office in Blagnac, France when they dreamt up the spare parts price list for Australia's MRH90s?

The Audit Office also reported that," In April 2012, on average each of the 15 in service MRH90 aircraft was costing approximately $51,200 per hour of flying, which Defence calculated to be 5.5 times more expensive than an ADF Black Hawk aircraft. At the same time, the cost of supporting the 15 MRH90 aircraft was more expensive than supporting the ADF's 34 Black Hawk aircraft." By the 2015 financial year

	Number	Cost ($m)	Hours Flown	Annual Cost per Aircraft ($m)	Cost per Flying Hour
Super Hornet	24	162	5,050	6.76	$32,080
AP-3c Orion	18	125	7,900	6.94	$15,820
F/A-18 Hornet	71	158	13,000	2.23	$12,150
Hawk LIF 127	33	91	7,500	2.76	$12,130
C-130J	12	98	7,350	8.17	$13,330
C-17	6	61	5,200	10.17	$11,730
MRH-90	47	157	5,400	3.34	$29,070
Seahawk	16	56	4,200	3.5	$13,330
Seahawk MH-60R	13	62	2,400	4.77	$25,830
Black Hawk	34	71	5,090	2.09	$13,950
ARH Tiger	22	114	4,726	5.18	$24,120
Wedgetail	6	163	3,600	27.17	$45,280
KC-30A Tankers	5	63	3,100	6.75	$32,080

the cost of operating the MRH90 was down to $29,070 per hour but still more than twice the cost of operating the Black Hawk.

Helicopters operating off ships have to be resistant to corrosion. The Defence Department's MRH90 Annual Structural Integrity Report, for the period 1st July 2012 to 30th June 2013, reported numerous instances of corrosion occurring throughout the MRH90. The study found that many components within the MRH90 aircraft corroded prior to delivery of the helicopters, and during the flight trial period. The MRH90 acquisition has been a complete dog. The Audit Office determined that "Assuming that MRH90 aircraft support (sustainment) costs may increase by three per cent per year due to price inflation, then the potential contracted services cost of sustaining the 47 MRH90 helicopters, in their current configuration, until their planned withdrawal date of 2040, may be in the order of $8.73 billion. On that basis, the total contracted cost of acquiring and sustaining the 47 MRH90 aircraft until 2040 will be some $11.7 billion."

What to do? The flyaway cost of a new Black Hawk UH-60M is US$17 million, made up of US$12 million for the airframe, US$0.662 million for the avionics package, two engines for US$1.5 million with other costs making up the remaining US$2.9 million. We could park up the MRH90s and replace them with new Black Hawks for $1,094 million. By 2028 the cost of buying and maintaining the new Black Hawks would be less than the cost of maintaining the MRH 90s if we had kept them. And we would have a helicopter fleet that was operable with our allies in this region. Selling off the MRH90s to European armies would further defray the cost of buying new Black Hawks. If Australia persists in keeping the MRH90s then we should copy what Poland is doing to keep their Soviet-era fighter aircraft flying. The Poles are now making parts out of plastic on a 3D printer for their Mig-29s, testing them for fit and then making permanent parts out of metal.

Australia has had six D-model Chinook helicopters which are being replaced by seven F-model Chinooks over the period 2014 to 2017. The Chinook weighs 9.7 tonnes empty, has a crew of four, a range of 600 kilometres and can carry 33 fully equipped troops. Chinooks, made by Boeing, cost about $50 million per copy.

Fixed wing aircraft

The Army doesn't have any fixed wing aircraft but it should. The Air Force is better at politics than either of the other two services. It got the Army to hand over the last of its fixed wing aircraft and then promptly degraded the capability. As the operation in the Solomons showed, RAAF logistics staff tend to unload cargo they don't think is important. The RAAF probably has enough airlift capacity for its own needs but not much more than that for the other services, and that's the way it is likely to work in practice.

Northern Australia is a vast place and Army commanders should have their own small aircraft that they can prioritise. An example of how things could and should be run in the United States Special Operations Command (SOCOM). SOCOM has a lot of freedom to acquire whatever equipment it sees fit. To move small groups of troops around the battlefield, SOCOM bought ten M-28 Skytruck aircraft from Polish manufacturer PZL. These can carry three tonnes of cargo or up to 18 passengers. The M28 is a westernised version of the 1960s era An-28 transport. It can cruise at 270 kilometres per hour for about five hours per sortie. Operations across northern Australian and into the Pacific Islands would soak up at least 60 Skytrucks. At US$4 million per copy, this capability could be had for about $300 million.

The alternatives would be to use scarce helicopters or for troops to drive hundreds or possibly thousands of kilometres. One of the sources of success on the battlefield is rapidity of movement. A good number of small transport aircraft will enable our forces to concentrate quickly. There would still be a need for more airlift capability for the Army.

Women in combat

There is one Western, liberal democracy that remains in constant danger of being overrun by its neighbours, neighbours who have been taught to hate it with their mother's milk. That nation is Israel. Israel has an interest in maximising the number of people who can be applied to combat so it trialled using women in combat. That wasn't successful for two reasons. Firstly, troops are trained to keep advancing despite the injuries and screams of their comrades who have been hit by enemy fire. If they don't, the attack might break and you end up with a worse situation – troops held down in exposed positions and the wounded and dying needing attention. Troops can be trained to happily ignore males screaming in pain but seem to be evolutionarily conditioned to go to the help of females in distress. This breaks the attack and makes such mixed units useless on the battlefield. Secondly, troops react to female comrades being wounded or killed by conducting reprisals against anyone to hand, enemy combatants or civilians. This isn't good for discipline. So Israel abandoned using women in combat because it just didn't work out. This is a great pity but you have to work with what you've got.

Complexities in modern warfare

Australian troops have been serving in Afghanistan since October 2001 as part of the futile attempt to do some nation-building in a place populated by violent, ungrateful wretches. Australian blood spilt there will have been for no purpose but at the same time we are grateful to those who served, and who will continue to be serving there into 2016 apparently. The burden of service has been particularly heavy on our special forces with some staff doing eight or nine tours of five months each in an eleven year period.

On 12th February 2009, two Australian soldiers from No 1 Commando Regiment were confronted by an armed Afghan in the doorway of his dwelling, which he withdrew into. The Australian soldiers threw grenades into the room with the Afghan which killed him and five of his children who were also in the room. On 27th September, 2010, the then Director of Military Prosecutions announced that manslaughter charges were being brought against the two soldiers and their commanding officer. The case against the three was dismissed by a military judge in May 2011. Of course the case was a figment of Director of Military Prosecutions' imagination. How her imagination works is illustrated by the fact that she had once described the treatment of David Hicks as "abominable". David Hicks is an Australian civilian who had gone to Pakistan to shoot at Indian troops across the border in Kashmir and went on to Afghanistan to serve with the Taliban in causing misery there. He had been captured in Afghanistan and then spent some time putting on weight in Guantanamo Bay.

As Orwell is quoted as saying, rough men stand ready to do violence so the rest of us can sleep peaceably. Some of those rough men, Australian special forces operating in Afghanistan in 2013, were in the news recently because they did their enemy dead the final respect of wanting to know who they had killed. So they collected some hands for fingerprint matching. An investigation of this incident followed. One of the troops involved has not yet been cleared after two and a half years. Now our rough men who protect us may be a bit hardened by their life experiences, but does anybody really think that they were souvenir hunting? That the trooper involved was going to post home a preserved hand for his trophy collection? An investigation of this sort of incident shouldn't take more than half an hour. Leaving that trooper in the process for two and a half years is an abuse.

Lasers Are Coming

It has been said that, in the future, for anything above the horizon to survive it will have to be highly reflective. The first operational laser deployed is on the *USS Ponce* which is based in the Persian Gulf. This is a US$40 million, 30 kilowatt laser which will destroy small scale threats to the *Ponce* such as small boats and drone aircraft. The tactics developed are that if the *Ponce* was threatened by a swarm of small boats, the laser beam would be held on one boat for the one to two seconds required to burn out a piece of equipment like the outboard motor before switching to the next one. Swarms of boats could be destroyed kilometres away from the *Ponce*. The laser can also disable aircraft and larger boats. The laser's cost per shot is US$0.59. The next stage for the US Navy is to deploy a 150 kilowatt laser that can shoot down incoming missiles. The US Air Force and US Army are also developing laser systems. China has announced an intention to build a fleet of 40,000 drone aircraft which would cost about US$10 billion. A cheap way to shoot those down is needed as soon as they appear over the radar horizon, and lasers will perform that role.

Commercial lasers for tasks such as welding use numbers of direct diodes, similar to an LED, to generate the laser light which is then concentrated using fibre optic filaments. The efficiency of conversion of electrical energy to laser light has reached 50 per cent. At that efficiency level, a five kilowatt laser will require 10 kilowatts of power. The other five kilowatts are converted to heat which is removed by refrigeration. A five kilowatt industrial laser sells for $300,000, rising to $600,000 for eight kilowatts. This is a unit the size of a refrigerator and would require a cooling unit twice its size. Industrial lasers have been coming down in price due to rising efficiency in semiconductor manufacturing. Efficiency in converting electrical energy to laser light has been going up at one per cent per annum. One big benefit from that is the reduction in the amount of cooling required and thus the size of the cooling equipment.

Deployment of battlefield lasers could be on vehicles the size of the Bushmaster with several vehicles networked to concentrate on one target at a time. Targeting information would be provided by ground-based

Figure 6: Bushmaster Protected Mobility Vehicle
The 12.4 tonne Bushmaster vehicle is the most successful design from Thales' Bendigo factory. The Australian Army and the Royal Australian Air Force have over 900 in service. Other countries that have taken the Bushmaster include Netherlands, Japan, Indonesia and Jamaica. Crews have had very high survivability in Afghanistan when attacked by improvised explosive devices.

surveillance radars such as the Saab Giraffe system. Australia has three of the Giraffe systems, bought to provide warning of incoming mortar rounds in Afghanistan. Ultimately lasers will be used to destroy incoming mortar rounds at the height of their ballistic arc of flight. But the biggest benefit of battlefield lasers will be to deny the enemy situational awareness provided by drones and aircraft. Australia could easily produce a truck-mounted battlefield laser and should gain experience and capability in the technology as a priority.

2

ROYAL AUSTRALIAN NAVY

History

Two months after federation of the Australian colonies to form the Commonwealth of Australia on 1st January, 1901, the naval forces of the separate colonies combined to form the Commonwealth Naval Forces. In 1908 it was decided that the force structure should be one battlecruiser, three light cruisers, six destroyers and three submarines. The first of these vessels, the destroyer *Yarra*, was acquired in 1910 and the following year the force was renamed as the Royal Australian Navy. World War I saw the Navy seizing German outposts in the southwest Pacific and then move on to the Mediterranean and North Seas. The Navy started the war with two submarines. The first sunk off Rabaul, New Guinea, without trace and the second was scuttled in the Sea of Marmara, Turkey, after a successful campaign.

The Navy shrank during the 1920s and 1930s due to apathy and cost pressures but then commenced growing again in the late 1930s with the approach of World War II. During that war the Navy became the fourth largest in the world, with 337 ships and almost 40,000 personnel. A total of 34 vessels were sunk during the war including three cruisers and four destroyers. Australian had no operational submarines during World War 2. Our first modern submarines were six *Oberon* class vessels starting from 1967 with the last decommissioned in 2000. The Navy also operated up to two small aircraft carriers with the last of these decommissioned in 1982.

Force Structure

The Navy basically consists of a surface fleet of 12 frigates and a subsurface fleet of six submarines. These are the vessels that will do the

Figure 7: HMAS *Anzac* in Sydney Harbour

The *Anzac* class frigates, first commissioned in 1996, are the major component of Australia's surface fleet. At 3,600 tonnes full load displacement, they are the ideal size for Australia's requirements. Ideally they will be replaced by a modern design with a low radar cross section, such as a slightly enlarged version of Singapore's *Formidable* class frigate which in turn is based on the *La Fayette* class from French shipbuilder DCNS.

bulk of the fighting There is some variation in type coming in that the surface fleet is being augmented by three destroyers, two large amphibious warfare ships and two replenishment ships being built in a South Korean yard. The Navy has two primary fleet bases. Fleet Base East located at HMAS *Kuttabul* in the Sydney suburb of Potts Point. Fleet Base West, at HMAS *Stirling* on Garden Island south of Perth, hosts the submarine fleet and five frigates. Patrol boats are based in Cairns and Darwin.

Technological advances have made the survival of surface vessels fraught and their utility moot. During World War II, for example, you might know your enemy's order of battle but most of the time you did not know where they were until you encountered them at sea, violently. These days satellites can track major surface combatants on a daily basis. If the satellites are put out of action, over-the-horizon radars such as Australia's JORN system can detect vessels up to 3,000 kilometres away. AWACS aircraft can detect vessels up to 700 kilometres away. Once detected, they can be targeted by anti-ship cruise missiles with ranges of up to 1,500 kilometres. Submarines can torpedo ships from up to 80 kilometres away, if they haven't hit them with cruise missiles first.

But a surface fleet is still needed for force projection and to be able to pick up survivors on the open ocean.

The submarine imperative

Australia can't be invaded if all sizable vessels approaching the Australian coastline are sunk. The only way to fully guarantee that is with submarines, if you have enough of them. Despite all the technological advances of the last twenty years, diesel-electric submarines remain undetectable, with no qualifier on undetectable. Allied diesel-electric submarines exercising with US naval forces regularly score hits on US ships. And so can unfriendly submarines. On 26th October 2006, a Chinese *Song* class attack submarine quietly surfaced within nine miles of the aircraft carrier *USS Kitty Hawk* as that 80,000 ton vessel sailed on a training exercise in the East China Sea between Japan and Taiwan. The *Song* class vessel, displacing 2,200 tons, was close enough to hit the *Kitty Hawk* with one of its 18 homing torpedoes. None of the carrier's roughly dozen escorting warships detected the Chinese submarine until it breached the surface. That incident was a shock to the US Navy.

An incident from the Falklands War illustrates how hard it is to sink a submarine that is stalking your forces. Early in the war the British nuclear submarine *Conqueror* sank the Argentinian cruiser *General Belgrano*, a former US World War II vessel with a crew of 1,138 officers and men. *Conqueror* fired three non-guided torpedoes, two of which hit the *General Belgrano* with the third hitting an Argentinian escort ship without exploding. Though the *General Belgrano* should have been "at action stations", she was sailing with the water-tight door open and 323 crew were killed in the sinking. The Argentinian surface fleet returned to its bases and did not venture out again during the conflict. At that stage the Argentinian Navy had one operational submarine, the *San Luis* built in Germany in 1973.

On 1st May, 1984, the *San Luis* fired a torpedo based on passive sonar detection of British gas-turbine-powered warships. The torpedo missed. It was heard by British sonar operators with the result that the British launched depth charge, mortar and torpedo attacks on contacts over the next 20 hours. All to no effect because the *San Luis* shut its systems down and rested on the sea floor, undetectable. The *San Luis* fired two more torpedoes on 10th May. One missed and the other did not leave its launch tube. After the war, German engineers went to Argentina to find out what went wrong with their torpedoes. What they found was that the Argentinian sailor in charge of periodic maintenance of the torpedoes had inadvertently reversed the polarity of power cables between the torpedoes and the submarine. The result was that when the torpedoes' gyros were spun up, they ran "backwards" and thus tumbled on launch, preventing the weapons from taking up their proper heading.

The British submarine counter-measures were equally as ineffective as the Argentinian submarine. The British fired over 150 pieces of ordnance to sink the *San Luis*, including 50 Mk 46 torpedoes. Including the two, ditched Sea King helicopters that the British abandoned due to the threat from the *San Luis*, the submarine's presence cost the British of the order of $300 million in current day dollars in capital equipment and ordnance. The *General Belgrano* was the second ship to be sunk by a submarine since World War II. The third was a South Korean corvette sunk by a North Korean mini-sub in 2010. That submarine escaped.

The submarine imperative is that submarines can be very effective at sinking enemy ships without being in much danger themselves. In comparison, surface ships can be detected at great distances and attacked from above and below at great distances. As former Defence Minister Kim Beazley said, "Basically submarines are the poor man's weapon to cause maximum angst to a bigger enemy", which describes the relationship between Australia and China. At the moment Australia is setting out to build eight frigates at a cost of $2.5 billion each, which happens to be slightly more than the cost of a *Virginia* class nuclear-powered submarine.

As a thought exercise, imagine a 200 ship-strong Chinese invasion fleet bearing down on the Australian coastline. The Chinese fleet is a mixture of frigates, oilers, helicopter carriers, amphibious landing ships and possibly an aircraft carrier. Standing in its way is Australia's fleet of eight new frigates. It would not matter whether or not we were able to get the first shot off. The Chinese would fire a salvo of perhaps 200 cruise missiles. At least fifty of those would get through and the Australian ships would be crippled or sinking. Any that were still emitting radar would get a second salvo. The Chinese fleet wouldn't even break stride in doing so. If that Chinese fleet faced eight *Virginia* class submarines instead, for the same capital cost to Australia, the Chinese would know that they would all be going to the bottom. With a submerged speed of 30 knots, *Virginia* class submarines could run down a surface fleet and engage and disengage at will.

How submarines operate

Submarines detect surface ships in three ways. The periscope is raised only a few centimetres above the sea surface so the maximum range that a target can be acquired visually is probably eight kilometres. The second way is from detecting the ship's radar or radio transmissions. This can be up to 200 kilometres if the ship is co-operating by transmitting on radar or radio. The third way is passive sonar if the enemy warship is transmitting on its sonar. This is effective up to 150 kilometres. Once again, the transmitting vessel can be detected far beyond the range that its active sonar is effective. Proponents of active sonar against submarines tend to forget what submariners call the "inverse fourth power" law. That is the power of the received signal that they get back as an echo of their transmission from the submarine they are hoping to detect is reduced in portion to the inverse of the fourth power of the range from the target submarine. The third way is for the submarine to use its passive sonar to detect the noise radiated from the ship itself, for example its engines, gear boxes and propellers. This is a more reliable source than any of the other means. The range of detection depends on the amount of noise being radiated by the ship and the environmental conditions in the area. A standard modern merchant ship is likely to be detected at about 40 kilometres.

A submarine's two main weapons are torpedoes and cruise missiles, launched through the torpedo tubes or from vertical launch tubes in the deck. Australia uses Mk 48 torpedoes which were originally developed in the 1970s to combat fast, deep-diving nuclear submarines and fast surface ships. With a range of over 50 kilometres at 40 knots, the torpedo can take an hour to reach its target. The torpedo can be heard by the target for the last quarter of an hour of that. It can be wire-guided all the way to the target or the torpedo can conduct its own programmed search and attack. It can also double back if it misses on the first attempt. Instead of penetrating the side of the ship, it explodes underneath it to break the ship's back. The latest version of the Mk48 torpedo is the Mod 7 Common Broadband Advanced Sonar System, jointly developed by the US Navy and the Royal Australian Navy. The Mk48 can dive to 900 metres. The purpose of that is to get under the thermocline, the boundary layer between warmer surface water and the colder water underneath, which tends to reflect and attenuate sound. This will decrease the distance over which the target ship can hear the torpedo coming and delay the use of countermeasures. In practice a torpedo might be fired at a depth of 70 metres and then dive deep to 600 metres before coming up to shallow depth again. The target ship can determine the bearing of the torpedo coming towards it but the ability of the torpedo to change its course mid-route means that the target cannot be sure of the bearing or range of the submarine attacking it.

The anti-ship cruise missile used in Australian submarines is a tube-launched version of the Harpoon. The Harpoon was so named because its original main target was to be surfaced Soviet submarines, nicknamed "whales". For submarine use, the Harpoon is encapsulated in a tube the size of a torpedo, fired from a torpedo tube and rises to the surface due to buoyancy. Upon breaking the surface, a cap blows off and a rocket motor fires followed by a turbojet burning jet fuel. The range is beyond 100 kilometres with a 221 kilogram warhead. If it is a lucky shot the Harpoon might disable or sink the vessel it hits. A US Navy study in the

1990s concluded that it would take five missiles on average to disable a large vessel, and seven to sink it. As a US admiral once said, missiles hit a ship above the waterline and let the air in, torpedoes hit a ship below the water line and let the water in. So cruise missiles, if they are substituting for torpedoes in the armory, decrease a submarine's effectiveness at sinking ships though they double its reach. They may also provide a radar track back to the submarine, betraying its location. The third type of weapon that Australian submarines carry is Stonefish mines which are pushed out of the torpedo tube and come to rest on the seabed.

While submarines can be used to launch cruise missiles at land targets, that is far more cheaply done, in terms of the capital cost of the platform, by long range aircraft. The cost of the Mk 48 torpedoes that Australia uses is US$3.8 million each. The capital cost of carrying that torpedo into battle by a *Virginia* class submarine is $100 million per torpedo. Anything that dilutes the primary mission of sinking major enemy combatants is wasting that capital outlay.

The history of submarines in Australia

Australia's first submarines were acquired in 1914. Two were purchased, the first of which sank off Rabaul later that year. The second was scuttled in Turkish waters the following year without loss of crew. While Australia did not have submarines (apart from one for sonar training) during World War II, Fremantle and Brisbane were important bases for American, Dutch and British submarines. The American submarines came from Manila as a consequence of the Japanese invasion of the Philippines. A couple of days after they arrived in Darwin, the Japanese bombed the port and the American force made its way to Fremantle. They were joined by Dutch submarines from the Dutch East Indies and later by British submarines transferred from the Atlantic. Approximately 160 individual submarines were based in Fremantle during World War II, conducting 416 war patrols. The American submarines of the time were originally termed fleet submarines. Based on doctrine developed from analysis of Word War I engagements, their role was to scout ahead of the main fleet and report the position of enemy combatants. To fit that role, the submarines were large with a displacement of 2,400 tonnes and a range of 20,000 kilometres.

The sinking of the US battle fleet at Pearl Harbor made fleet combatant role redundant, freeing them to become commerce raiders. After problems with malfunctioning torpedoes were rectified, their range and endurance made them quite successful. By comparison, the British and Dutch submarines based out of Fremantle sometimes had to come back from patrols early because their crews had developed sores from working in the fetid air of un-air conditioned craft operating in the tropics. They also had barely enough range to get to their patrol areas before they had to return to base. The American submarines had air conditioning which, among other benefits, stopped condensation from affecting the electrics. A fueling station in Exmouth Gulf, 1,100 kilometres north of Fremantle, extended the operating range by 2,200 kilometres. Brisbane was the other major submarine base during that war with Cid Harbour in the Whitsunday Islands as a fueling station. From the 1960s, Australia operated six *Oberon* class submarines built in the UK. These were a direct descendant of the German Type 27 submarine with a submerged displacement of 2,400 tonnes and two diesel engines. Planning to replace the *Oberons* began in the 1970s. From the seven submitted proposals, two were selected for a funded study to determine the winning design which would become the *Collins* class – a German one from HDW and a Swedish one from Kockums. The German offer was made in early 1987. It was expected that the German design would win the competition with the Swedish bid in to keep the Germans honest. The Labor Government at the time though gave the order to the Swedes and this became the *Collins* class programme.

The Swedish design was an enlarged version of a submarine designed for the comparative puddle of the Baltic Sea. For example, the initial refrigerator capacity was good for a two week deployment, not ten weeks. The *Collins* class had a troubled beginning on several fronts. The initial consortium included Chicago Bridge and Iron (CBI), Kockums and Wormald. CBI realised early on that the Swedes don't do scheduling and so the project would drift. CBI offered to buy out Kockums and take

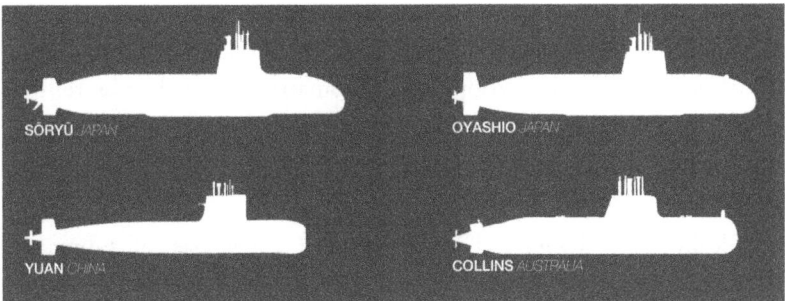

Figure 8: Profiles of the *Soryu, Oyashio, Yuan* and *Collins* Class Submarines[1]

The *Soryu* class derived from *Oyashio* class, first of which was commissioned in 1998. The main differences between the two classes of vessel is that the *Soryu* class has X-shaped tail planes, air-independent propulsion and weighs 150 tonnes more at surface. The next vessel in the *Soryu* class will have lithium ion batteries instead of air-independent propulsion. China has approximately 15 of the *Yuan* class in service, as well as *Song* class, *Kilo* class and *Ming* class submarines. The *Yuan* class has a submerged displacement of 3,600 tonnes. The *Soryu* class has a surface displacement of 2,900 tonnes and a submerged displacement of 4,200 tonnes which is 1,300 tonnes greater. The *Collins* class has a surface displacement of 3,100 tonnes and a submerged displacement of 3,407 tonnes which is 307 tonnes greater. The difference in submerged tonnage between the two classes means that the volume of *Soryu* class is 23 per cent larger than that of the *Collins* class. This means that it takes more power to push it through the water but at the same time the *Soryu* class is inherently safer. If a *Collins* class submarine was damaged or blew a hose, such as the incident on *HMAS Dechaineux* on 12th February, 2003, and took on 308 tonnes of water, it would not be able to surface. The *Soryu* class could take a lot more damage and water ingress, and still blow its tanks to get to the surface. The Japanese submarine builders would have a lot more institutional memory of operating submarines in wartime than the Swedish designers of the *Collins* class.

over the running of the project. Instead Kockums bought out CBI and then encountered financial difficulties due to the end of the Cold War and the resultant decline in defence spending. Kockums was subsumed into a Swedish consortium called Celcius, onsold to HDW, which in turn was bought by ThyssenKrupp. At the same time Australia started with the *Collins* class, the UK had a similar experience with its *Upholder* class, four of which were commissioned between 1990 and 1994. The first three built were unable to fire torpedoes, despite the UK's long history of designing, building and operating submarines, so had to be refitted. Then, with the end of the Cold War, the UK decided to go to an all-nuclear fleet and the four *Upholder* class submarines were sold to Canada at a bargain basement price.

Ideally in the design and construction of a new class of submarine, the first of the class is launched and operated for a while before laying the keel of the second. This is to remove most of the bugs early in the programme. Unfortunately the second hull in the *Collins* programme was well under construction when the first one was launched. The result was that all the *Collins* class required costly refitting instead of mainly just the first one. In essence, the *Collins* class was very casually designed by the Swedes, matched by a lack of interest by the Navy who weren't aware how bad the *Collins* class was until they were put in the water. For example, a report by Lloyds Register observed, "The number of diesel engine-related failures sustained onboard HMAS *Farncomb* and indeed the *Collins* class submarines as a group is totally unacceptable to the point of being quite unique in modern medium speed diesel marine applications." To power the *Collins* class, Kockums chose a 12 cylinder stationary diesel produced by the small firm of Hedemora, increasing its power by making an 18 cylinder version. The Navy could have and should have changed out these defective diesels during refitting but instead is persevering with them to the bitter end. We will have the *Collins* class for at least a decade yet and the diesels should be changed out at the earliest opportunity.

Collins class submarines are best summed up in the words of a former submarine commander, James Harrap, who retired from the Navy in 2012. It is worth quoting his words, from Asia-Pacific Defence Reporter of 4th May, 2012, in full:

> The maritime patrol aircraft was still out there – an Australian AP3C with the formidable ELM2022 radar, entirely unforgiving to submarines in most circumstances but for some reason not today. I instructed the watchleader to remain at periscope depth since it gave us the best chance to identify and intercept our target, the

approaching task group consisting of frigates HMAS *Newcastle* and HMNZS *Te Kaha* and the tanker HMNZS *Endeavour*. Our assessment of the oceanography was such that periscope depth was also favourable for avoiding detection by the ships' sonar. There was of course the chance of visual detection by the aircraft too, with the submarine only being five metres below the waves, or the periscope – raised for 30 seconds every six minutes – leaving a wake if we tried to go too fast. But with a southerly wind at 20 knots there were plenty of white caps around and in my experience the benefit of remaining at periscope depth was worth the risk. Over the next 90 minutes the tense scenario played out. First the frigates then the tanker became visual. The AP3C seemed to be obsessed with tracking something further to the north – possibly a whale or just some disturbance in the ocean – that worked in our favour. We got into the position I wanted, about 7 kilometres away with good bearing separation on all the targets, ideal for our torpedoes. 'Pass the Deepfield message [indicating a torpedo attack] then let them know where we are and get the hell out of here' were my instructions to the watchleader, dutifully carried out. HMAS *Collins* dived to 100 metres, increased speed and headed off, north, back to where the aircraft was until recently searching – I thought it unlikely it would revisit the old datum since it was proven false.

This was March 2012 and HMAS *Collins* was in the Western Australian Exercise Areas conducting Exercise Triton Storm, an annual high-end training event. It was my last three weeks at sea as a submarine captain and there couldn't have been a better way to finish off. Operating at a heightened but sustainable degree of readiness, the crew bring together skills, equipment and experience to deliver a capability feared and admired throughout the world and desperately needed as part of Australia's national defence. It's just getting it all together which is the big challenge.

There are a few times in each of our lives when we make a complete career change and set course in an entirely different direction. For me now is one of those times, having completed almost 20 years in the Royal Australian Navy, 15 of them as part of the submarine force, culminating in Command of submarines HMAS *Waller* and HMAS *Collins*.

Leaving the navy direct from command is uncommon, though not unprecedented; I am however the only command-qualified submariner in several years to depart for a non-defence and non-government related industry. Whilst my reasons for the sudden change are many and varied, this article is about my experience as a *Collins* Class submariner. Since 1999 I have served onboard all six *Collins* Class submarines and intend to provide an insider's view on one of the most intensely debated and captivating defence projects of the last two decades.

My submarine experience has spanned the full spectrum of submarine operations and I have sailed on six different classes of submarines (three nuclear-powered and three diesel-electric) with five different navies. I don't pretend to be an expert on all aspects of submarine design and operations, but I consider myself to have a pretty good idea. One thing common to all my submarine experience is that no design is perfect and no single navy has the monopoly on superior equipment and procedures. Military and naval capability is not, as many believe, resident only in the specifications of the hardware employed. Capability is a much more complex equation depending on: weapons, equipment, personnel, communications, command infrastructure, training and experience to name but a few. In the same way that being a winning Formula 1 racing team is not just about the car, a submarine is not an advanced naval capability just because of the platform. Informed commentators and military professionals understand this; unfortunately journalists and politicians often do not. Consequently, much of what has been written about the *Collins* Class, and proposed replacement options I believe to be inaccurate or incomplete.

Contrary to the perception held by many, the *Collins* Class has served the RAN well and achieved much. Despite some well-publicised failures, the successes of the *Collins* Class are also numerous and have been significant though, as is the nature of submarine operations, these have not always been widely publicised. However, as I became more experienced and learnt my trade as a submariner it was seldom concerns about any foreign enemy which occupied my time, but rather fighting the enemy within. Submarines are highly complicated machines and being a submariner has always required a skilled blend of operator/technician unique within naval service; but the *Collins* Class has taken the technical arguments to a whole new level. The planned maintenance requirements are onerous enough but the constant stream of defects and operation control limitations makes getting to sea difficult, staying at sea harder and fighting the enemy a luxury only available once the first two have been overcome. The submarines have maintained an operational capability for most of the past 15 years, but that is often despite many aspects of the submarine's design rather than because of it.

Numerous limitations and failings of the *Collins* Class have been made public but somehow, with support from government and key private contractors, the Navy has managed to work through or around most of these. I have no doubt that this can continue to be the case for the next few years, but there are now some disturbing themes which must be addressed. I can't help but think that, just like me, it is perhaps time for a drastic change here too.

Reliability, sustainability and crewing constraints have been persistent concerns in recent years. The links between these three issues compound the challenge faced by the submarine force. Reduced reliability and increased failure rates on equipment place a greater demand for replacement parts and labour thus increasing sustainment costs whilst also limiting the number of available submarine sea days. This lack of available sea time limits training and experience opportunities for the crews and slows the rate at which the submarine force can re-grow.

Sustainment and reliability problems have plagued the *Collins* Class since first launch. This is not news to anybody who has followed the history of the boats. As commissioning crew onboard HMAS *Rankin*, I recall fitted equipment being 'cannibalized' to support the other boats even before the submarine was commissioned. The expected reply to stores demand signals is usually 'Nil Stock Global' reflecting no suitable inventory holdings – though this is often attributable to accounting inadequacies as much as it is a true reflection on the state of inventory. Over the last two years though I believe these problems have become worse; throughout my command of both *Collins* and *Waller* full capability was never available and frequently over 50 per cent of the identified defects were awaiting stores. Sustainment is a very topical issue and costs are quoted in the order of $400-500 million per year (and total operating expense about $800m) with various organisations currently working on managing this problem. During 2011 the Navy instigated a Submarine Logistics Continuous Improvement Program to address known inadequacies in submarine logistics support. Additionally the Coles review was initiated by the government in 2011 and will cover some similar ground. Whilst these initiatives and others are probably justified, and no doubt will result in some improvements, they will not provide a miracle cure to the problem. Lack of available stores inventory, increased equipment failure rates and submarines living with reduced capability is something I expect will persist for the remaining life of the Class.

In 2008 the Navy's submarine workforce numbers fell to an almost unrecoverable level due to significant attrition of qualified personnel and lack of recruitment to this volunteer force. Re-growth of the submarine workforce is critical to maintaining and building submarine capability for the Navy. The seagoing workforce currently consists of three submarine crews with a desire to stand up a fourth as soon as practicable, each crew consists of about 60 officers and sailors of various skill sets and experience levels. The submarine force is supported ashore by additional personnel in various essential roles such as logistics, maintenance, operational planning and training to name just a few. Some of these positions do not require seagoing submarine experience, but the bulk of direct support roles do. Training and experience growth is consistently a top priority for the Navy but lack of available submarine sea days reduces opportunities to build not just the seagoing but also the support workforce. I believe the Navy is currently doing a good job to recover the submarine workforce, though unnecessarily encumbered by the inadequacy of the defence personnel management framework. The situation is fragile and will remain so for some time to come.

Of equal importance to the 'uniformed' navy workforce is the support provided by the Defence Materiel Organisation, ASC, other prime contractors and government organisations. As my time in the submarine force progressed I have come to know many of the key support personnel on the waterfront, many of them former submariners with whom I have previously served at sea. The Navy long ago gave up the ability to conduct all maintenance itself and external support is essential; so whilst not always considered as part of the submarine force numbers, the people within these organisations are vital to sustainment of submarine capability too. Many of these people are dedicated and capable individuals - though I can't say the same for all components of the support organisations. Skills shortages here also impact on submarine maintenance schedules, work quality, availability and ultimately capability.

Failure to adequately address the human element will lead to a demise of submarine capability just as rapidly as an inability to put submarines to sea caused by materiel deficiencies. The future of submarine capability must consider both the navy and non-navy components of the personnel dimension.

Another aspect of sustainment is the need for capability enhancement in order to take advantage of new technology. When I commanded HMAS *Waller* in 2011, it was a different boat to the one I received my dolphins onboard in 1999. The combat system and torpedo were entirely different, the sonar was similar but much improved and other more subtle changes such as the sewage system and ISCMMS (the boat's platform control and monitoring system) were evident. The *Collins* Class have evolved but the pace of change is slow and the cost is high.

Sustainment budgets and schedules must continue to factor the requirement to improve the capability. Much of the existing equipment is bespoke (and often obsolete), the need for upgrades is increasing but the cost of acquiring and retrofitting equipment is high. Rising numbers of defects swell work scope during maintenance periods and merely getting the scheduled maintenance done in the allocated time is a challenge before capability replacement and upgrade is even considered.

Whilst continuous improvement is essential, there comes a time when the incremental changes possible in this process are not enough. Some components of the submarine are either not able to be changed or to do so would carry a prohibitive mix of risk and cost. The *Collins* Class has many components that we are simply stuck with for the life of the platform. For example the diesel generators fit into this category because of their size; unfortunately they are quite possibly the least reliable diesel engines ever built. They have been problematic throughout the life of the class and, despite some design modifications and improvements, are only kept running by ingenuity and sheer determination of the crews at sea and supporting contractors alongside. Because of components and immutable design issues such as these, *Collins* has a finite service life.

There is also a component of 'technology pull' that limits the effective life of any submarine platform. To extend the earlier analogy: it would be impossible to win next year's grand prix in a 20 year old car – no matter how good the driver and support team are. HMAS *Collins* when first launched was hailed as being ahead of its time; I don't entirely agree with that, but it was the vanguard of a new generation of submarine design. That was in the early 1990s. Since then numerous advances have occurred in batteries, electric motors, air-independent propulsion, sonars and electro-optics – all of which have revolutionised submarine design even further. These changes have been significant and whilst it may be possible (though very costly) to keep Collins operational for another decade or more, most advances can't be retrofitted and the boat will most likely be so technically obsolete by 2022 that the credibility of the capability it offers will be seriously eroded.

Despite the problems I have highlighted above, I still standby my earlier comment that our submarines deliver a significant capability and that this is because of the whole package, not just the platform but all other components as well. This could not happen without a solid commitment and strong leadership by government and the most senior levels of Defence to sustain the capability. The skill and resolve to do this is admirable but doesn't come without having made some difficult choices and shown the tenacity to see them through. This is definitely a theme which must continue. To ignore the decision on the future of the submarine force is to choose to erode any advantage we still possess, because within our region submarine capability is increasing at an alarming pace.

China continues to build submarines at a rate unmatched anywhere in the world whilst the quality and capability of the Chinese submarine fleet increases faster than the nation's GDP. India has this year taken delivery of a Russian-built *Akula* SSN and continues with its own construction programs. Other regional nations are buying or building their own submarines at a ferocious pace, all with recently developed technology unavailable onboard a *Collins* Class. I am not advocating we join a new arms race, I am saying we have been in one for quite a while and we need to keep up. The 2008 Defence white paper made a clear case of the continued requirement for submarines and while the type and numbers may be debatable the need for the capability is not.

There are a number of options available when it comes to our future submarine force. If my time onboard *Collins* has taught me anything it is that any submarine is better than no submarine at all, so my first recommendation is simple – do something. I don't believe that the *Collins* Class are sustainable in the long term and many of the expensive upgrade plans which have been proposed would be throwing good money after bad. Though sustaining what we currently have is essential until we can get a replacement class of submarine commissioned. Growing the size and experience of the submarine force requires boats at sea; that must be our primary aim. Secondly, reliability and long-term sustainability are crucial for our future submarine. Lack of platform reliability is the single most limiting factor for the *Collins* Class, let's never repeat that mistake. A submarine capable of most of the tasking available most of the time is better than one that claims to do all of the tasking but is only available some of the time. For any future submarine, robust through-life support and sustainment is just as important as any other design specification.

Australia's strategic circumstances are unique, but so are those of Singapore, India, Japan and every other nation. That does not mean that our next submarine platform needs to be entirely unique though. As I have stated, the Australian submarine capability is vested in more than just the platform, so even with an existing design the capability would remain unique and would be superior to other similar platforms if supported in superior ways. For example, if we are able to retain the Mk48 Mod 7 torpedo (in my mind the best torpedo available) we could operate a similar submarine to our adversary and still have a clear advantage. Additionally, a cheaper initial build and sufficient funds for through-life upgrade and sustainment is better than a more expensive platform which we can't afford to service or upgrade. We don't need the best submarine money can buy but rather the best submarine capability we can afford.

Many of the arguments supporting the unique requirements for our future submarine focus on long duration patrols, extended ranges, and lengthy covert ocean transits. Whilst a scenario can be created to necessitate this, you can't let one extreme and hypothetical situation define the reality of our future. I do not believe a diesel-electric significantly larger than *Collins* is possible, much less a good idea. There will always be some missions that can't be achieved, let's focus our solution on the ones which can.

Whatever decision is made regarding our future submarine, it will remain with us for many years to come. The value of a modern and capable submarine is significant, but so are the costs of ownership and we must be realistic about what can and can't be achieved. *Collins* has consistently been let down by some fundamental design flaws leading to poor reliability and inconsistent performance. This has taken a toll on submarine availability and sustainability of our workforce. The cost and requirements of through life support and capability upgrades were poorly planned and have been difficult to implement. Whilst we have been able to fight on with the *Collins* Class, the challenge of doing so has been significant and will continue to increase. An inability to keep up with rapid technological change, coupled with high materiel failure rates has aged the boats prematurely, adding cost and complexity to through-life support. The boats must be sustained in the short term, but I do not believe a service life extension for *Collins* is even possible, much less recommended.

The navy has continued to struggle with crewing and supporting its submarines, as have Defence Materiel Organisation, ASC and others, we must strive for best practice as we continue to address personnel issues. I do not believe we have the capability to independently design and build our own submarines, nor do we have the ability to grow the submarine workforce at a faster pace than what has been achieved over the past two years. Our future must build upon our past, with regard for our failings as well as successes. We can't afford to set ourselves impossible requirements, because we will surely fail to meet them."

So sayeth Commander Harrap. Indeed, operating and maintaining the *Collins* class submarines is costing Australia $100 million per vessel per annum. By comparison, the nuclear-powered *Virginia* class submarines,

Figure 9: HMAS *Rankin*
Rankin was the last of the 3,000 tonne *Collins* class submarines. She was laid down on 12th May 1995, launched on 7th November 2001, delivered to the Royal Australian Navy on 18th March 2003. Commissioning on 29th March 2003 was three and a half years behind schedule after major delays in the completion and fitting out of the boat due to fast-tracking of *Dechaineux* and *Sheean* and repeated cannibalisation for parts to repair the other five *Collins* class boats.

at 7,900 tonnes weighing more than twice as much as the *Collins* class submarines and more than twice as capable, cost the American taxpayers US$50 million per annum to operate. Australia should replace its *Collins* class submarines as soon as we can. We also need a lot more than six. Fortunately, like the modern woman, we don't have to choose and be denied anything. We can get both diesel-electric and nuclear-powered submarines directly off their production lines. The best choice for Australia is the *Soryu* diesel-electric submarine from Japan and the *Virginia* class nuclear-powered submarine from the United States. Both countries are willing and able to sell them. And we need a few more bases to put them in.

Soryu Class

Japanese submarine construction began in 1904 when they bought some small submarines in kit form from the United States. Japanese technology got a leg up after World War I when they were given some German submarines. Come World War II and Japan was producing a wide range of submarines, from the midgets that penetrated Sydney Harbour, to giant 6,500 tonne vessels that carried three seaplanes within them. One of the latter sent a seaplane to scout over Sydney, Hobart and Melbourne during the war. Japanese submarines sank 18 ships off the Australian coast and damaged another 25, with 467 killed. For the sake of completeness, a German submarine shelled a tanker 200 kilometres south of Adelaide and also sank vessels off Jervis Bay and Fremantle. One amusing incident is that a Japanese submarine shelled Port Gregory, on the West Australian coast 440 kilometres north of Perth, as a diversion for an operation they were conducting in the Pacific. The Australian Government wasn't aware that Port Gregory had been shelled until a Japanese message was decrypted a week later.

Following World War II, Japan recommenced building submarines in 1959 and has built a further nine classes of submarine since then. Basically Japan has been continuously building submarines since 1959. They keep their submarines for 16 to 18 years after which they are scrapped. That is they take their submarines through one deep maintenance cycle but avoid the cost of a second one, preferring the updated technology of a new submarine build. Japan is in the ideal situation of a continuous design and build process, incorporating improvements into the next submarine to come off the production line rather than the risks involved in a completely new design. For example the current *Soryu* class started with Air Independent Propulsion (fueled by diesel and liquid oxygen) to supplement the lead acid batteries. The next boat in the class will switch to lithium ion batteries which have twice the energy density of lead acid batteries and do away with the Air Independ Propulsion. Another advantage of continuous build is that the Japanese have their costs under control. The next boat has been given a budget of US$537 million, and doubtless the quality is faultless.

The US Navy exercises with many other navies and came to the conclusion that the *Soryu* class is the best diesel-electric submarine available. It has been criticised for its stated range of 11,000 kilometres but large portion of the vessel was taken up by the Air Independent Propulsion system. Now that that is gone, there is a lot of space freed up for more diesel storage. There is another important reason for opting for the *Soryu* class without much further consideration. Japan could boost its production rate of the class to perhaps two a year and sell Australia one a year. At that rate we might be up to a full complement of 12 by 2028 and the *Collins* class could be completely retired by early next decade. If we made any other choice, the first boat might not be operational until the mid-2020s and we wouldn't have the full complement until the 2030s. Australia's strategic situation compels us to get as many submarines as we can as soon as possible.

The Federal Government has announced a $20 billion frigate building programme to be conducted in South Australia. Hopefully that will provide enough clear air politically to proceed to a contract for supply of submarines from Japan. There is not much point in pronouncing how many submarines we should get because our plans, whatever they are, will be overtaken by events and proven to be inadequate. We should acquire as many as we can, as soon as possible. A total of 24, equating to one per million head of population, might be appropriate but we wouldn't reach that level by late next decade at the earliest. There are some 300 submarines in the Asian region. A count of 24 would give us about 8 per cent of the total of the region which would be just enough to have an influence. Bear in mind that if we had 24, a dozen might be available for patrol at any one time for patrol in the vastness of the Indian Ocean and the western Pacific. Of those dozen, half might be on station and half may be transiting to or from their patrol areas.

Virginia Class

The United States began building nuclear-powered submarines in 1952 and stopped producing diesel-electric ones in 1959. By the end of the Cold War, the United States had three classes of submarine – the *Los Angeles* class for attacking other submarines and shipping, the *Ohio* class for carrying ballistic missiles and the *Seawolf* class, which is also an attack class. But the *Seawolf* was too expensive to build so production halted at three. The *Virginia* class was developed to replace the *Seawolf* as a more affordable vessel. The first hull was laid down in 2000 and eleven have been commissioned to date. The full production run may be 50 boats with some remaining in service to 2070. The reactor of the *Virginia* class does not need refueling over its 33 year life, which also sets the life of the submarine.

Two are built each year as this is cheaper than building at one annually. Two shipyards are involved. The Huntington Ingalls shipyard in Newport News, Virginia, builds the stern, living quarters, machinery spaces, torpedo room, sail and bow. The Electric Boat shipyard in Groton, Connecticut, builds the engine room and control room. The yards alternate work on the reactor plant as well as the final assembly, test, outfit and delivery.

The downside of nuclear power is that the reactor can't be completely switched off. Decay of fission products in the reactor continue after all the control rods have been inserted and so pumps need to keep circulating water to keep the reactor from heating up. This means that nuclear-powered submarines will never be as quiet as diesel electric ones and thus are less suited to the shallow waters of parts of Southeast Asia. Their upside is that they can travel indefinitely at 30 knots underwater with their range only limited by the food carried on board. The *Virginias* carry 27 torpedoes, three less than the *Soryu* class, but also have 12 Tomahawk cruise missiles in vertical launch tubes.

Australia should acquire nuclear-powered submarines for the same reason as the United States. We are surrounded by vast areas of open ocean which we need to traverse quickly to concentrate forces against an enemy ship concentration. The faster our submarines can travel, the larger the area enemy forces at are risk in, circumscribing what they might attempt to do.

If for some reason we cannot obtain the *Virginia* class, we should opt for the French *Barracuda* class submarine. France is building six of these at a cost of €1,300 million (A$2,000 million) per unit. The nuclear reactor on the *Barracuda* class has a ten year fueling cycle but has a crew of only 60 in a vessel with a surface displacement of 4,765 tonnes.

Basing Considerations

Australia's current submarine fleet of six are based in HMAS *Stirling* on Garden Island south of Perth. If you go to 32° 13' 41" S, 115° 41' 33" on Google Earth, you can see two of them tied up at the submarine wharf. This is a problem in that HMAS *Stirling* is within range of cruise missiles carried by Chinese bombers with inflight refueling, and certainly by Chinese submarines launching short range missiles just off the Western Australian coast. Chinese military doctrine holds that a war should be started with a surprise attack so the first indication of being at war with China might be the sinking of our submarines and frigates in port at HMAS *Stirling*. Part of the power of submarines is that if the enemy doesn't know where they are then it has to account for the fact that they could be anywhere. At the moment Chinese reconnaissance satellites can count the submarines tied up at HMAS *Stirling* on a daily basis. At the minimum we should build overhead cover for the submarines to keep them guessing.

During World War II, the US Navy operated a fueling base codenamed Potshot at Exmouth Gulf, halfway up the West Australian coast. Topping up at Potshot gave Allied submarines from Fremantle more time on patrol in the South China Sea. The same holds true today. The range of diesel-electric submarines depends upon their tonnage. Being able to refuel at Exmouth would add range equivalent to 1,000 tonnes of displacement. With the *Soryu* class costing $800 million for about 3,000 tonnes of surfaced displacement, the effect is equivalent to about $250 million per submarine. Over 12 submarines, that would amount to some $3 billion worth. The necessary basing infrastructure, overhead cover included, could be built for a fraction of that. Exmouth Gulf is a good anchorage with a large area with 14 metres depth of water. It should be backed up with a similar base at Shark Bay to the south.

Australia is three thousand kilometers across which is a problem for submarine operations in the Pacific. The *Collins* class submarines were being worn out just by the 5,000 kilometer round trip to Adelaide and back to be repaired. Supposing a submarine was directed to support a task force heading out into the Pacific from Townsville. It would take two weeks to get to Townsville from HMAS *Stirling*. Australia's enemies aren't going to cooperate by restricting their operations to Indian Ocean and South China Sea. The solution is to have a submarine base on the east coast, preferably in North Queensland to minimise the time to get to the operational area. The ideal location is Double Bay on the Whitsunday coast, 10 kilometres north of Airlie Beach. During World War 2, submarines based in Brisbane topped up at a base in Cid Harbour on the western side of Whitsunday Island, just across the channel from Double Bay, so the location has precedent. Another precedent from World War 2 was that another fueling base was maintained at Manus Island on the northern side of New Guinea, further extending range.

Surface Fleet

As pointed out at the beginning of this chapter, the survival of surface warships in the modern age is problematic. If they emit sonar while searching for submarines, they are alerting submarines up to 150

Figure 10: *Formidable* Class Frigate RSS *Steadfast*
The Singaporean Navy *Formidable* class frigate is derived from the *La Fayette* design by the French shipbuilder DCNS. It combines a low radar cross section with a minimum crew number of 71, with a further 19 crew for the helicopter carried.

kilometres away of their presence. From that distance, a diesel-electric submarine would take a couple of hours to get into a launch position for a torpedo shot and the torpedo itself would take another hour to reach its target. Or the submarine might just launch a cruise missile which would reach the ship in seven minutes. If warships use radar to track aircraft 200 kilometres away, they are alerting aircraft twice that distance away of their position. An anti-ship cruise missile would take less than half an hour to get to our radar-emitting warship from 400 kilometres away. Even if a warship doesn't emit sound, radar or radio waves, it is still a big infrared target on the background of the cold open ocean. That said, we still need a surface navy which we should make as survivable as possible.

The warfighting surface fleet Australia has at the moment consists of 12 frigates. To put that into context, military surface vessels start in size from patrol boats of few hundred tonnes, corvettes around 2,000 tonnes, frigates start from 3,000 tonnes, destroyers from 6,000 tonnes and cruisers from 9,000 tonnes. A US Navy study found that cost-effectiveness in ocean-going warships started at 3,000 tonnes. Below that, vessels did not provide value for money. Also, the bigger the size, the rougher the seas that a boat could handle.

Our frigate fleet is made up of four *Adelaide* class ships of 4,100 tonnes displacement and eight *Anzac* class ships of 3,600 tonnes displacement. The *Adelaide* class is based on the US *Oliver Perry* class. Of the six built, two have been decommissioned to date. The remaining four were commissioned between 1983 and 1993 and so are also due for replacement in the next few years. The *Anzac* class frigates are based on the German MEKO 200 design and were commissioned between 1996 and 2006.

These frigates are being supplemented by another two classes of warship. Three *Hobart* class destroyers, based on the Spanish *Alvaro de Bazan* class, will be added to the fleet from 2017 to 2019. The *Hobart* class, of 6,250 tonnes each, are optimised on air defence against aircraft and missiles. The second class of warship to be added is also based on a Spanish design, in this case the *Juan Carlos 1* class amphibious assault ship. In Australian service these will be two *Canberra* class Landing Helicopter Dock ships. The *Canberra* class are big ships with displacement of 27,500 tonnes at full load. They were ordered in response to the trauma the Australian military went through in trying to supply forces ashore in East Timor in 1999. That was an unopposed landing 600 kilometres from Darwin that almost failed due to inadequate logistics.

That fleet is what we have and are getting. The question is, if we started with a blank sheet of paper, what should we have? Some things can be done more cost-effectively by other types of platforms. Submarines are a more cost-effective way of sinking ships than using ships to do that. Aircraft are a more cost-effective and versatile way of delivering anti-ship cruise missiles than ships are. Those aren't problems, it just means that our surface vessels are relieved of the demand to devote a lot of their resources to those things to the detriment of what their role should be. That role is to provide persistent presence and be good at anti-submarine warfare. We don't need a large ship to do that. The ideal vessel for our requirements is the Singaporean *Formidable* class frigate of 3,200 tonnes. Australia would need longer range which would be achieved by adding 10 metres to the length of the hull and taking displacement to 3,500 tonnes. The *Formidable* class has very good radar cross section reduction with inclined hull sides and bulwarks. Due to having a lot of automated systems, its manning requirement is 71 versus about twice that for the *Anzac* class. Singapore, with a population of six million, has six of them which it built for about US$600 million each. Australia's need is a lot greater than the current fleet of 12 frigates. As well as the three *Hobart* class destroyers, a fleet of 21 frigates would take us to a surface fleet of 24 warships.

The *Hobart* class will be with us for at least a couple of decades but their philosophy of operation and thus design is wrong. To counter the threat from aircraft and missiles, they are sized to carry powerful

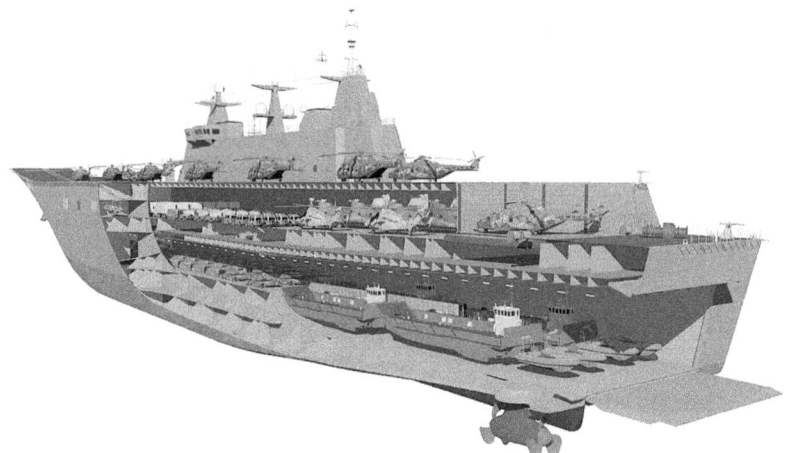

Figure 11: Cutaway View of the *Canberra* Class Amphibious Assault Ship
The *Canberra* class is a 20,000 tonne amphibious assault ship with a displacement of 27,500 tonnes at full load. The *Canberra* and its sister ship, the *Adelaide*, will carry six helicopters as standard. The ships have accommodation for 1,046 soldiers and will carry four landing craft with the ability to take tanks ashore.

radars and the long range Standard missile. Hostile aircraft and missiles can be detected a long range from the ship and intercepted a long way from it. The Standard is an expensive missile, costing $3 to $6 million depending upon the variant. If it is used to shoot down cruise missiles costing $0.5 to $1.0 million per copy, our *Hobart* class destroyers will run out of Standard missiles long before the enemy has exhausted his supply of cruise missiles. Each vertical launch cell holding one Standard missile could hold four of the Evolved Sea Sparrow missile instead. The latter missile has a 30 kilometres range and is a fraction of the price of the Standard missile.

The *Canberra* class amphibious assault ships went the other way with almost no protection against anything - floating, submerged or airborne threats. They will be equipped with four 25 mm, remotely-controlled guns and six 50 calibre machine guns which could take on speedboat-sized threats. To be deployed in battle, the *Canberra* class will be accompanied by at least five or six frigates which will try to stop cruise missiles from passing through the fleet to hit the *Canberra* or its sister ship, the *Adelaide*. Some will leak through though and it was a mistake to not provide the *Canberra* class with any anti-air protection. By comparison, the US *Wasp* class amphibious assault ships, of about 50 per cent greater displacement than the *Canberra* class, have four anti-air missile stations and two Phalanx close-in-weapon-system stations. Not providing anti-air systems to the *Canberra* class was probably a cost-saving measure. Effectively the *Canberra* class will be used for police work in the Pacific nations. History suggests though that they will at some stage be used in a contested environment and the false economy of not having any anti-air protection may have tragic consequences. Providing protection to both HMAS *Canberra* and HMAS *Adelaide* would take all of our current fleet of frigates.

While Australia's acquisition of the *Canberra* class was in response to the logistics strain of the East Timor exercise in 1999, the US military's response was in the opposite direction. The Navy had leased an aluminium catamaran from its builder Incat in Tasmania and commissioned it as HMAS *Jervis Bay* on 10th June 1999. The 1,250 tonne vessel could cruise at 45 knots. The lease was an experiment to fill a capability gap. In 1994, the Navy had spent $60 million buying two ex-US Navy tank landing ships, the USS *Saginaw* and USS *Fairfax County*, which became the HMAS *Kanimbla* and HMAS *Manoora* respectively. The ships were then found to require $400 million of repairs. The East Timor crisis began in September 1999 and HMAS *Jervis Bay* proved very useful in providing rapid transport from Darwin, a fact noted by US observers.

That observation resulted in three classes of ships in US service. In 2001, the US Marine Corps chartered the 2,111 tonne *MV Westpac Express* built by Austal Ships at Henderson, south of Perth. The *Westpac Express* remains under charter to the US Marine Corps. In 2003, the US Navy chartered the 1,668 tonne *HSV-2 Swift*, built by Incat in Tasmania, to trial high speed catamarans as a test platform. This experience led to the trimaran-hulled *Independence* class of littoral combat ships built by Austal in Alabama and the *Spearhead* class of military sealift vessels.

There had been a notion that the *Canberra* class could be equipped

Figure 12: HMAS *Armidale* Patrol Boat
The *Armidale* class patrol boats are 300 tonnes displacement with a top speed of 25 knots and a range of 5,600 kilometres at 12 knots. They are virtually unarmed with one 25 mm Bushmaster autocannon in a Rafael Typhoon stabilised gun mount on the forward deck and two 50 calibre machine guns, though the 20,000 tonne *Canberra* class amphibious assault ships don't carry much more. The range of the 25 mm cannon is three kilometres.

with the short-take-off-vertical-landing version of the F-35, the F-35B to be used by the US Marine Corps. Apart from that aircraft's ineffectual performance, its cost is horrendous at US$251 million per copy. Two F-35Bs would pay for another frigate. The *Canberra* class also has no way of providing fire support to troops ashore. The *Anzac* class frigates have a 130 mm gun with a range of 24 kilometres. A faster, more effective way of providing fire support with longer range would be to use the HIMARS GPS-guided multiple rocket system. This is an eleven tonne truck carrying a pod of six missiles, each carrying a 91 kg warhead. The HIMARS system has a range of between 2 and 85 kilometres. The unit cost of the truck and one pod of missiles is US$4.2 million. They could be kept on the light vehicle deck of the *Canberra* class vessels and brought up to the main deck using the forward elevator.

Other Vessel Classes

The Navy also has patrol boats, minehunter, survey ships and logistic support ships. There is a notion, floated with the building of frigates in South Australia, that the first three classes be replaced by one hull type of up to 2,000 tonnes called an Offshore Combatant Vessel. This will not work because form follows function and the resultant ships will be too big for their tasks. The US Navy followed a similar philosophy with its corvette-sized littoral combat ships of 2,300 tonnes displacement. They have a large internal volume for mission modules but a lot of that space is wasted. At the same time they are lightly armed with the consequence that they are not expected to be survivable in a hostile combat environment.

There is a political element to the Federal Government's decision to build the next generation of frigates in South Australia. That might work well if good managers run a productive workforce making a good design. An example of how naval shipbuilding as an employment exercise can go wrong is provided by the UK's two *Queen Elizabeth* class aircraft carriers of 70,000 tonnes displacement. The project was started by then Prime Minister Gordon Brown to provide work in Scottish shipyards. Originally costed at £3.9 billion, the projected cost of the project is now

£6.2 billion ($13 billion). Each is supposed to have a complement of 12 F-35B fighter aircraft plus helicopters. At US$251 million each, the F-35Bs would cost $3.0 billion per ship. The *Queen Elizabeth* class aren't fitted with catapult-assisted take-off and wire arrestor gear. Fitting those things would cost £2.0 billion per ship.

Although the F-35 is capable of landing vertically, this places limitations on the loads that the aircraft is able to return to the ship with. As a consequence, to avoid the costly disposal of fuel and bombs at sea, the Royal Navy has developed the shipborne rolling vertical landing technique for operating the F-35Bs on the *Queen Elizabeth* class. This is a hybrid landing technique that uses vectored thrust to slow forward speed to around 70 knots to make a rolling landing, using its disc brakes, without the need of an arrestor wire. The 2010 UK Strategic Defence and Security Review declared that the UK only needed one aircraft carrier. The penalty clauses in the construction contract meant though that cancelling the second ship would be more expensive than actually building it. The review recommended that the second ship would be mothballed on completion or sold.

Actually using an aircraft carrier means that it has to be protected by six to ten, frigate to cruiser-sized warships, an oiler and usually a submarine. The UK doesn't have the ships to devote to that role, even if the F-35Bs are delivered. The likely fate of both vessels is that they will be sold and used as helicopter carriers.

The current patrol boat in service in the Royal Australian Navy is the *Armidale* class built by Austal at Henderson, south of Perth. Of the 14 built, 13 remain in service after HMAS *Bundaberg* caught fire while on the hardstand in Brisbane. The *Armidales* have a displacement of 300 tonnes and cost $28 million per ship. Using the experience of the *Armidales* in service, Austal built eight *Cape* class patrol boats for the Australian Customs and Border Protection Service. For the same tonnage displacement, the *Cape* class have 30 per cent larger internal volume for a two metre increase in length. Though the *Armidale* class boats are only ten years old, they have been heavily used and need replacement. The logical development would be to build military versions of the *Cape* class as Australia's next patrol boat. Currently our patrol boats are based in Cairns and Darwin. There are no naval vessels stationed along the 3,400 kilometres of coastline from Darwin to Perth. In particular, the 2,000 kilometre stretch from Darwin to Exmouth faces the region where Australia will do the bulk of its future naval and air warfare. Australia could do with a larger patrol boat fleet with some of them based in Exmouth.

3

ROYAL AUSTRALIAN AIR FORCE

History

The Royal Australian Air Force (RAAF) originates from a decision at a conference of British Empire countries in 1911 to develop an aviation branch of the armed services. The Australian Government approved the formation of a Central Flying School in 1912. The first flights of the school, located at Point Cook in Victoria, took place in March, 1914 and trained pilots for the Australian Flying Corps. The Corps saw action in Egypt, Palestine, Mesopotamia and the Western Front in the First World War. This unit was disbanded in 1919. It was succeeded by the Australian Air Corps in 1920. This in turn was separated from the Army in 1921 as a separate service called the Royal Australian Air Force. Upon formation, the RAAF had 170 aircraft and 149 personnel.

The RAAF was expanded in the late 1930s in anticipation of the Second World War, from under 1,000 personnel in 1935 to around 3,500 in 1939. By the end of the War, the RAAF had become the fourth-largest air force in the world with over 152,000 personnel operating nearly 6,000 aircraft. Australia's population at the time was less than one third the current level. A total of 216,900 Australians had served in the RAAF during the war of whom 11,061 were killed in action. A high proportion of these combat fatalities occurred in crews attached to Royal Air Force Bomber Command. Two per cent of all RAAF personnel served with Bomber Command during the war but accounted for 23 per cent of all RAAF fatalities. For example, No. 460 Squadron RAAF, flying Avro Lancasters, had an official establishment of about 200 aircrew and had 1,018 combat deaths. The squadron was effectively wiped out five times over.

The RAAF remained active after the Second World War and

participated in the Korean War, Malayan Emergency, Vietnam War, both Gulf Wars and the East Timor Campaign. The RAAF currently has eight Super Hornets based in the United Arab Emirates to battle Islamic State. The force supporting the Super Hornets has 400 personnel and includes a KC-30A tanker (based on the Airbus A330) and a E-7A Wedgetail aircraft (based on the Boeing 737). As at 2014, the RAAF had 13,991 permanent fulltime personnel and 4,316 part-time active reserve personnel.

Fighter aircraft introduction

Australia cannot be invaded if at least either of two conditions hold – that Australian submarines sink any invasion fleet approaching the Australian coast; or Australian fighter aircraft maintain air superiority over Australia and out to the Indonesian archipelago. The latter condition would mean that enemy surface forces, on sea or land, could be interdicted at will. Ideally we would maintain both capabilities to be sure and make the whole job easier. If we don't maintain air superiority over northern Australia and its approaches then life, and staying alive, becomes far more difficult for the rest of our armed forces. So having the right fighter aircraft in the requisite quantity to achieve air superiority is one of the two major considerations in our force structure.

The first fighter jet in Australian service was the de Havilland Vampire with 110 built by the Commonwealth Aircraft Corporation in Melbourne, which also made its engine. The second was the F-86 Sabre produced by North American Aviation. It was introduced in 1956 and retired in 1971. The Sabre was followed by the Mirage III in 1964. Designed by Dassault Aviation of France, it was built by the Government Aircraft Factory at Fisherman's Bend in Melbourne. The Mirage III was capable of Mach 2.2 and had a higher top speed than the aircraft that replaced it, the F/A-18 Hornet. Australia had some 116 Mirage IIIs. Of these, 50 were sold to Pakistan and that country still has some of our former Mirage IIIs in service.

The "A" in F/A-18 is short for attack and means that the aircraft was designed to perform ground attack as well as provide air superiority. The genesis of the aircraft was a design competition for a lightweight fighter for the US Air Force in the early 1970s. The US Air Force chose the single-engined YF-16 which became the F-16 Fighting Falcon in service. The US Navy chose the aircraft that lost the Air Force competition, the twin-engine YF-17, because it did not want to have the same aircraft as the Air Force and because it wanted a two-engine aircraft for over-water flights. This became the F/A-18 Hornet in service. Australia ordered 75 of them to replace the Mirage IIIs. Of those, 72 are still flying. One of the reasons that Australia chose the F/A-18 over the F-16 was that it was built more ruggedly for aircraft carrier landings. The F/A-18 has a maximum speed at 40,000 feet of Mach 1.8.

From the 1970s Australia also had the F-111 in service. While it had "F" for fighter in its designation, this swing-wing aircraft was purely a light bomber with the capability for dropping nuclear bombs. That said, the F-111 had a lot of capability in that it had a combat range of 2,100 kilometres while being able to carry 14.3 tonnes of bombs on an aircraft with an empty weight of 21.4 tonnes. By comparison, the swing-wing Tupolev Tu-22 "Backfire" bomber carries up to 24 tonnes on an aircraft that weighs 58 tonnes empty with only a slightly longer range.

By the beginning of the 21st century, the F/A-18 was becoming outdated and outclassed by more recent Russian fighter aircraft acquired by Indonesia and Malaysia. It was also designed for an airframe life of 6,000 hours. As the replacement for the F/A-18, Australia plumped for the F-35 while it was still in development. That was over a decade ago and the F-35 is still in development. As a stop-gap measure, Australia bought 24 F/A-18F Super Hornets with deliveries beginning in 2010. This was followed by an order for an additional 12 Super Hornets in the electronic warfare variant.

Australia's initial commitment was for 14 F-35s though only two were contracted for. The Abbott Government added another 58 to take the total commitment to 72 which is the number of F/A-18 Hornets to be retired. The F-35 is one of the worst choices in the history of Australian military procurement. It is incapable of air superiority against

Figure 13: Sukhoi Su-35
China is in the process of acquiring 24 Sukhoi Su-35s from Russia. This is the most advanced variant of the Su-27 Flanker that first flew in 1977. China's illegal copies of the Su-27 are the Shenyang J-11 and Shenyang J-16, of which it has close to 250. In battle simulations of the F-35 against the Su-35, 2.4 F-35s are lost for each Su-35 shot down. If that ratio is borne out in practice, the 24 Su-35s in Chinese service will account for 58 F-35s.

contemporary and future threats like the Sukhoi Su-35S, T-50 and Chendu J-20 and is outclassed by the F-16, first produced in the 1970s. Australia's reliance on the F-35 puts our security at risk. But the aircraft is so bad that production is likely to stop before we get most of what we ordered. That is a good thing. It also means that we have to find a replacement fighter aircraft. The sooner we have that process in train, the safer and more secure we will be. The right choice for Australia's next fighter aircraft will be the Gripen E from Saab in Sweden.

The evolution of fighter aircraft

The first jet fighter aircraft, the Messerschmitt ME-262, took to the air on 8th April, 1941, and became operational in 1944. The first Allied jet, the Gloster E.28/39, had its first flight five weeks later on 15th May, 1941. The major designs of the 1950s were the US F-86 Sabre and the Soviet Mig-15. These were single-engine aircraft weighing 6.9 tonnes and 6.1 tonnes respectively. By the 1960s, size increased to 8.8 tonnes for the Mig-21 with a US equivalent, the F-104 Starfighter, weighing 9.4 tonnes. Then the US built the F-15, a twin-engined aircraft optimised around its large radar and designed primarily for high level interception of Soviet bombers. The F-15, still in production after nearly fifty years, has a loaded weight of 20.2 tonnes. In those days, ability to detect enemy aircraft depended upon the size of the radar which was mounted in the fighter's nose. The further away they could be detected, the greater the advantage to the fighter aircraft which could then launch beyond-visual-range, radar-guided missiles. So as the size of the radar grew, the size of the aircraft had to grow with it. The F-15 also had a gun because one of the major lessons of the Vietnam War was that most missiles missed. No F-15 has been lost in combat and it has 104 kills to its air combat record.

The Soviet response to the F-15 was the Sukhoi Su-27 with a maximum take-off weight of 30.4 tonnes. The trend continued up to the F-22 Raptor which has a maximum take-off weight of 38.0 tonnes. That is more than five times the weight of the Sabre and more than the empty weight of the B-29 bomber of World War 2 of 33.8 tonnes. Apart from housing a large radar, the design philosophy of the F-22 is to have a small radar cross section to avoid detection by enemy radars, either on the ground or in other aircraft and to be able to 'supercruise' at Mach 1.7+ and operate above 60,000 feet.

At the same time the F-15 was being designed, a group in the US Air Force nicknamed the Fighter Mafia realised that air superiority would be more cost effectively achieved by a small, single-engine fighter that was highly manoeuvrable with a high thrust-to-weight ratio. This concept bore fruit as the F-16, also still in production after nearly fifty years. It has a loaded weight of 12.0 tonnes and a maximum take-off weight of 19.2 tonnes, half that of the F-22.

Russian and Chinese design efforts have followed the lead set by the F-22. The first Russian stealth fighter is the T-50, weighing an estimated 35 tonnes at maximum take-off weight. The first Chinese stealth fighter, the J-20, is slightly heavier at a 36.3 tonne maximum take-off weight. All these designs are so expensive that not many are going to be built. In fact production of the F-22 stopped at 187. Two have crashed so there are only 185 still flying. The F-22 costs US$250 million each to build. That is one thing, but they are also quite expensive to fly at US$50,000 per hour of flight. They are also maintenance-intensive with 40 hours of maintenance for each hour of flight. In turn that means that they have a low availability and a low sortie rate. F-22 fighters are so expensive to operate that the pilots don't get enough monthly hours to be properly proficient in operating them. As pilot skill is a large part of air superiority, this negates in part the F-22's advantages. Costs of operating the Russian and Chinese stealth aircraft will be much the same.

Fortunately for Australia, technological developments have swung the air superiority pendulum back towards the lightweight, highly manoeuvrable single-engine fighter.

Fighter design considerations

Fighter aircraft should be hard to detect and highly manoeuvrable in order to surprise and outmanoeuvre the enemy as well as to improve survivability against missile fire. To achieve that requires small size, supercruise ability, good aerodynamic design, low wing loading and high thrust-to-weight ratio. Supercruise is the ability to maintain a speed above Mach 1.0 without the use of the aircraft's afterburner. Wing loading is the loaded weight of the aircraft divided by the area of the wing. The aircraft that uses its radar first will be quickly detected and targeted by passive sensors. Therefore only minor radar cross section-reduction measures are needed.

Low observability (being hard to detect) and sensor fusion (consolidating the aircraft's sensor inputs) are required to achieve the advantage, getting off the first shot and possibly achieving a kill with a low chance of being targeted in return. If that doesn't work, breaking the enemy's OODA loop (the Observation, Orientation, Decision, Action loop concept developed by John Boyd) by being impossible to predict is essential. The ability to supercruise helps in both as it shrinks enemy's response time after the supercruiser is detected, reduces effectiveness of the opponent's weapons while increasing effectiveness of the supercruiser's weapons, allows the supercruiser to achieve surprise while preventing the enemy from surprising him, and to dictate terms of engagement.

Manoeuverability is important in air combat for two reasons: to get the enemy inside one's own engagement envelope, and to avoid getting hit. While some modern fighters such as the Rafale and the F-35 can use missiles to engage aircraft directly behind them, this is of questionable usefulness as it increases target's reaction time and causes the missile to lose energy, as well as increasing the likelihood of missile simply not acquiring the target. It used to be that missiles used in beyond-visual-range, having spent their fuel and flying on inertia alone, would have a low chance of hitting a manoeuvring target. The solution to that problem that the US adopted is the 'two pulse' motor of the AIM-120D. But this makes it too fast to turn the corner at the terminal kill and thus it still has a high chance of missing. The European missile maker MBDA developed its Meteor missile to throttle back from Mach 4 to below Mach 2 for the terminal kill and as a result can turn into a target turning at 9G at 50,000 feet.

Manoeuverability in a fighter aircraft requires the ability to start turning quickly and then to have a high sustained rate of turn. But the most important requirement is the transient performance – that is roll onset, turn onset and pitch rates as well as acceleration, deceleration and instantaneous turn rate. This needs high lift-to-weight, lift-to-drag, thrust-to-weight and thrust-to-drag ratios while sustaining high g (and consequently high angles of attack) as well as generally low drag at all speeds and high control power with ability to generate large amounts of drag when required. The instantaneous turn rate in particular needs low wing loading and a high lift coefficient. Maximum turn rate and minimum turn radius is experienced at an aircraft's corner speed; for the same g limit, a lower wing loading results in a lower corner speed and thus a higher turn rate and smaller turn radius.

The best way to achieve these characteristics in a fighter aircraft is a blended wing-body configuration with a delta wing and close-coupled canards positioned in front of and high above the wing. The blended wing-body configuration achieves greater lift and lift-to-drag values than conventional configurations such as the F-15 and increases the available volume inside the aircraft. It also reduces the radar cross section and wave drag from the formation of shock waves in supersonic and transonic flight. This is why there are now three European delta wing/canard combinations – the Dassault Rafale, the Eurofighter Typhoon and the Saab Gripen. When the Israelis set out to build their own fighter aircraft, that effort produced a delta wing/canard fighter called the Lavi. Similarly, when China produced its first modern jet fighter it was a delta wing/canard combination called the J-10.

The total lift of the close-coupled canard configuration is far higher than the additive lift of the wings and the canards. This is a result of their beneficial interference when in close proximity, with the canard acting like a 'forward flap". This enhancement can be effective to such extent that maximum lift is 34 per cent greater for a close-coupled canard configuration than for an otherwise identical configuration with no canard, with canard adding only 15 per cent of the area. Canards also increase the angle the aircraft can fly at without stalling.

A canard mounted above the wing has a noticeably better lift-to-

Figure 14: Saab Gripen (Copyright Saab AB, photo credit Peter Liander)
This photograph is of the C variant. The distinguishing marks of the E variant are an infrared search and track housing ahead of the canopy along with a cooling-air scoop at the base of the tailfin.

drag ratio than a coplanar canard, as the vortex and wake-flow from the canard do not hit the wing. Maximum lift is achieved when the canard's trailing edge is slightly in front of the wing leading edge. Moving the canard forward or down reduces the lift gain. A properly positioned canard creates a low pressure region on the front part of the wing upper surface which has a significant contribution to lift.

Launcher rails on the wing tips allow two missiles to be carried with virtually no drag penalty while improving the lift-to-drag ratio. The body of a fighter aircraft should be slightly wasp-waisted in order to comply with the area rule, but it is not a necessity as modern jet engines are powerful enough to push even a flying brick like the F-35 through the transonic region. The area rule is based on the fact that at high-subsonic flight speeds, the local speed of the airflow can reach the speed of sound where the flow accelerates around the aircraft body and wings. The speed at which this development occurs varies from aircraft to aircraft and is known as the critical Mach number. The resulting shock waves formed at these points of sonic flow can greatly reduce power which is experienced by the aircraft as a sudden and very powerful drag, called wave drag.

To reduce wave drag the cross sectional area of the aircraft should remain as constant as possible down its length and changes in cross sectional area should be as smooth as possible. Thus the fuselage should be narrowed where the wings are attached to account for the cross sectional area of the wings so that the total area does not change much. Nevertheless a fighter aircraft should not spend much time in the transonic region as it should be either cruising or manoeuvring at supersonic speeds or manoeuvring at subsonic speeds. Of modern fighters, only the Gripen and the F-18 use the area rule. The Typhoon, F-22, F-35, F-15 and F-16 all ignore it though the F-35's combination of a relatively low thrust and high transonic drag results in abysmal acceleration characteristics.

Fighters are built with one engine or two. Twin-engine air superiority fighters have higher survivability than single-engine fighters. For example, the single-engine F-16 has twice the number of engine-related mishaps of the twin-engine F-15. The reason why Australia still has 72 'Classic'

Hornets is that we have had many engine failures and the aircraft have returned on one engine. On the other hand, single-engine fighters are more manoeuvrable, especially in roll and changing direction, and so are better able to avoid getting hit in the first place. That said, the vectored thrust of the twin-engine F-22 and Su-30MK increase their manoeuvrability. In the Sukhoi's case, one engine can be vectored up and the other one down to increase roll rate. Single-engine fighters have smaller visual and infrared signatures.

A single engine helps reduce cost in several ways. Single-engined fighters are more amenable to area ruling, which means that they tend to have less drag and thus lower fuel consumption. This can also lead to reduced size, weight and thus procurement cost as well. Maintenance downtime required is also lower. All of this leads to single-engined fighters having significantly lower direct operating cost than twin-engined fighters. Thus the operating cost of the F-16C is US$7,000 per hour versus US$16,500 per hour for the twin-engined Rafale C. The Rafale C has an 11 per cent greater empty weight and 28 per cent more dry thrust than the F-16C yet costs 2.4 times as much to operate. The F-15C costs US$30,000 per hour to operate, yet has a 48 per cent higher empty weight and 52 per cent more dry thrust compared to the F-16C. More complex aircraft also require more maintenance personnel: the Gripen needs 10 assigned flightline maintenance personnel, compared to 30 for the F-15.

Situational awareness is one of the most important characteristics of an air superiority fighter. This starts with visibility from the cockpit and is improved with a variety of sensors. Cockpit visibility is divided into two basic sectors: forward visibility, required for early target detection, and an aft visibility, which is crucial for avoiding an attack from behind. The pilot also has to be able to visually check for threats in the rear quadrant, and also to see whether or not the aircraft is producing any contrails.

At beyond visual range, onboard sensors are crucial in detecting and identifying other aircraft. Radar cannot reliably identify the detected aircraft and it warns them of the scanning aircraft's presence far before it actually can detect them, thus allowing them to take measures appropriate for the situation.

Unique radar characteristics enable enemy aircraft to identify the fighter using the radar, and the radar itself is vulnerable to electronic countermeasures. Modern anti-radiation missiles also enable fighters to passively target the emitting aircraft. Identification-friend-or-foe will be kept off as it allows the enemy to track the fighter. Thus the most important sensor for an air superiority fighter is the infrared-search-and-track sensor as it can detect and identify faraway targets completely passively – up to 70 kilometres in good conditions but not in cloud. Radar warning receivers are also important but they depend upon enemy aircraft using their own radars which is not likely to happen in a war.

With respect to weapons, the main missile type used should be infrared-guided. Radar-guided missiles are easy to counter and are thus ineffective. They need 15 seconds to lock on, allowing ample time for the radar warning receiver to detect and analyse the attacker's radar emissions. Secondary beyond-visual-range missiles should have a combined radar-homing and infrared seeker in order to provide diversity in seeker types.

In the Vietnam War, probability of kill was 26 per cent for the aircraft's gun, 15 per cent for the Sidewinder missile (within-visual-range with an infrared seeker), 11 per cent for the Falcon missile (beyond-visual-range with an infrared seeker) and 8 per cent for the Sparrow missile (beyond-visual-range with a radar receiver). During that war, 51 kills were made with guns, 83 with heat-seeking missiles and 56 with radar-guided missiles. In the Yom Kippur and Bekaa Valley wars, Israel made 93 kills with guns, 225 with infrared missiles and 17 with radar-guided missiles (two at beyond-visual-range). It can be seen that infrared, within-visual-range missiles are a fighter aircraft's primary weapon, and opportunity for engagement depends on identifying the enemy – usually visually.

In the First Gulf War, radar-guided missiles achieved a kill probability of 27.3 per cent, indicating that missile reliability had not improved much since the Vietnam War. There were only five confirmed beyond-visual-range kills in the First Gulf War, despite radar-guided missiles accounting for 24 kills out of 85. F-15s performed far better than other Allied fighter types with a radar-guided kill probability of 34 per cent - 23 kills out of 67 shots, and an infrared missile kill probability of 67 per cent - 8 kills out of 12 shots. By comparison, the US Navy's F-14s and F-18s achieved a radar-guided kill probability of 4.8 per cent – one kill out of 21 shots, and an infrared kill probability of 5.3 per cent – two kills out of 38 shots. In the Second Gulf War, F-16s fired 36 Sidewinder missiles for zero kills, though 20 of these launches were accidental due to poor ergonomics of the control stick.

In terms of kill probability, guns have a kill probability of between 26 per cent and 31 per cent, infrared within-visual-range missiles of 15 per cent, infrared beyond-visual-range missiles of 11 per cent and radar-homing, beyond-visual-range missiles of 8 per cent. The kill probability of beyond-visual-range missiles falls by 25 per cent compared to values listed when they are actually used at beyond-visual-range.

Traditionally, heat-seeking missiles required five to seven seconds to lock on, obtain parameters and launch compared to 10 to 15 seconds for radar-guided missiles. The pilot would have to point the nose of the aircraft at the target to obtain a lock. The development of the helmet-mounted cueing system and high-angle, off-boresight missiles has reduced these times. This combination was developed by South Africa for their war against Angola. The seeker in the missile head follows where the pilot is looking by tracking the position of the pilot's helmet in the cockpit. The pilot only has to look at the target and fire the missile, which will lock-on after launch. The Soviet Union noted the success of the South Africans in shooting down the Soviet-supplied aircraft and copied the technology. When East Germany was reunited with West Germany, the West found out how effective the Soviet technology had become.

A gun kill requires three to six seconds. Seven seconds is the maximum safe time for achieving a kill during a dogfight. A fighter in a dogfight shouldn't keep the same course for more than seven seconds. Otherwise enemy fighters will be figuring out how to attack it.

The Eurofighter Typhoon's infrared-search-and-track sensor can detect subsonic fighters at 90 kilometres from the front and at 145 kilometres from the rear. The jet engines themselves are very hot and they heat up the airframe surrounding it. Apart from the engines and their exhaust, they are a number of other sources of infrared radiation from an aircraft. Movement of the aircraft through the air leads to compression of the air in front of it. This heats the air. For example a super-cruising aircraft at Mach 1.7 generates shock cones with a temperature of 87°C. Friction from the air heats the aircraft's skin. In a jet fighter, the hottest parts apart from the engine nozzles are the tip of the nose, front of the canopy and the leading edges of the wings, tail and engine intakes. Modern infrared-search-and-track systems can detect missile launch from nose cone heating.

Unlike radar, infrared-search-and-track is primarily a passive system. This allows a fighter aircraft, or a fighter group, to detect and track the enemy without latter being aware of their presence, thus gaining a significant initial advantage. Even when the enemy is aware of the fighter's presence, he has no way of knowing whether or not he has been detected, or is being targeted, until a significant shift in the fighters' posture, such as painting a target with a rangefinder or shifting flight path or formation. For comparison, just turning on the radar warns the aircraft in a very large area of the scanning fighter's presence – and the said area is far larger than one covered by the radar. Not only does it give away fighter's presence, but if the enemy has good-enough listening equipment, it is possible to triangulate the location and even identify the target through its unique radar signals. Even radio communications and datalinks can serve the same purpose.

If the enemy is using radar, it is possible to use data from a radar-warning-receiver to generate a bearing, after which infrared-search-and-track can be used in a "stare" mode – continuous track, during which photon impacts are combined over prolonged timeframe to detect a target at greater distances than would normally be possible. This mode is also present in radar systems, and like infrared-search-and-track, radar also has to be cued by other sensors to make use of it. But while using radar in such a manner basically guarantees that the enemy with a competent radar-warning-receiver will detect radar transmissions, infrared-search-and-track is undetectable. Even a short radar burst can allow the passive fighter to generate a bearing.

If radars are jammed, or more likely turned off for fear of detection, the first indication of an infrared-search-and-track equipped fighter's presence that the enemy aircraft will get may be the alarm from its missile warning system, thus allowing only a short time for defensive reaction. If both sides have infrared-search-and-track, it comes down to sensor quality and infrared signature differences.

Aircraft equipped with infrared-search-and-track, and using an infrared missile approach warning system, can remain completely silent during the mission. If the enemy has no infrared-search-and-track, then he will have to turn on his own radar, allowing the passive aircraft excellent situational awareness, well beyond what using radar in addition to infrared-search-and-track would allow. Radar is not the primary onboard sensor anymore and is not actually even required.

The latest variant of the Gripen, the E model, uses an infrared-search-and-track system called Skyward G. This sensor weighs 30 kg. It is a dual-band system covering the midwave and longwave infrared bands, and can provide an infrared image on the pilot's visor. Scan coverage is 160° in the horizontal plane and 60° in elevation.

Skyward G is stated to be capable of detecting all aircraft flying faster than 300-400 knots from skin friction alone – irrespective of any exhaust plume or engine infrared signature. It can track more than 200 targets simultaneously.

The F-22 does not have an infrared-search-and-track system, which means that it has to use radar to engage the enemy at beyond-visual-range. It was dropped as a cost-saving measure on a US$250 million aircraft. This, combined with its large size and high infrared signature, severely limits its ability to achieve surprise bounces. In terms of avoiding surprise it is no better. While limited rearward visibility is somewhat compensated for by the high cruise speed of Mach 1.7, its high infrared signature despite some infrared signature reduction measures means that it will be easily noticed.

If the enemy uses very-high-frequency and high-frequency radars, the value of stealth is heavily reduced if not eliminated altogether – as shown by the F-117 shot down over Serbia only 18 seconds after getting discovered by the very-high-frequency radar, and another F-117 that got mission-killed by the same surface-to-air missile battery. The latter F-117 returned to base but was damaged beyond repair.

The Russian T-50 appears to be optimised to shoot down US fighter aircraft, primarily the F-22 and F-15. China's J-20 is more optimised for shooting down US airborne-warning-and-control aircraft, transport and tanker aircraft, thus neutralising relatively short-range US fighters without having to engage them in combat at all. The F-22 is a compromise between two roles. The J-20 is meant to avoid aerial combat though it should be able to handle itself if it comes to that.

If up against a good pilot in a superior fighter, one can win if the opponent is forced to make a mistake. For this, one must be a better pilot than the opponent – and good pilots are made largely by in-flight combat training as opposed to simulator training. This means that ease of maintenance, reliability and low operating costs are important characteristics of a fighter aircraft if pilots are to get enough flight time to be proficient. Today's US Air Force F-22, F-35 and F-16 pilots get 8-10 hours of flight training per month, and US Navy pilots get 11 hours per month. French Rafale pilots get 15 hours per month, while RAF Typhoon pilots get slightly more at around 17.5 hours per month. This can be compared to a minimum of 20-30 hours per month required for fighter pilot to be truly proficient.

The number of missiles carried also determines fighter effectiveness. The more missiles carried, the more that can be fired in a salvo. Russian Su-27s fire a two, three or four missile salvo. Kill probability of a two missile beyond-visual-range salvo is 19 per cent, of a three missile salvo 27 per cent and of a four missile salvo 34 per cent. The rate of kill also depends upon the time to solve a firing solution.

Over the last five years, Gripen, Rafale and Typhoon fighters have come to the Red Flag air combat exercises in Alaska to be matched up against US aircraft. With their advanced electronic warfare suites and superior data links, the Gripen and Rafale fighters had no problem in locating the F-22s and remaining undetected by everything but powerful radar scans, which would have led to the destruction of the radar-emitting aircraft by the radar-homing Meteor missile. F-22 pilots reported these smaller fighters were upon them within-visual-range before the F-22's vaunted electronics suite could detect them. The smaller Gripen was within gun fighting range before being detected. F-22 pilots were forced to go vertical to escape most of the time using their huge Pratt and Whitney engines at full afterburner which is not a good technique against even an average pilot with heat-seeking missiles. The 2015 Red Flag Alaska was highlighted by one German Typhoon recording three kills against F-22s. And that is with the Typhoon not being allowed to use its infrared-search-and-track under the rules of the engagement.

Small size is important for avoiding detection by high frequency sky-wave and surface-wave radars. Sky-wave radars, such as Australia's JORN system, bounce their radar waves off the ionosphere. Surface-wave radars also use high frequencies from 3 MHz up to 30 MHz. Electromagnetic waves at this frequency tend to bend or diffract around edges or curves. They are coupled to the conductive ocean surface forming a "ground wave", bending over the horizon and following the curvature of the earth. The Gripen's resonant frequency is about 26 MHz which is rarely used in military radars. Bigger aircraft like the F-35, F-22, B2, J-20 and T-50 have resonant frequencies in the 10-15 MHz range - the sweet spot of high frequency over-the-horizon radar.

The F-35

It has been said that the story of the F-35 begins in 1942 in the Battle of Guadalcanal. The US Marines, doing the ground fighting, were upset that the other services weren't providing enough air cover. The pounding they got from the lack of air cover is part of their institutional memory. So when the US Defense Department decided to build a 5th generation stealth fighter to replace the F-16, the US Marines insisted that this include a short take-off and vertical landing (STOVL) variant. The trade-offs necessary to effect this fatally compromised the whole project so that none of the variants do their job adequately. Specifically, the requirement to have a lift fan 1.27 metres in diameter on the centreline of the aircraft behind the pilot resulted in two bomb bays instead of just

Figure 15: F-35A
The F-35 started as a design for a light bomber that would come through after F-22 fighter aircraft performed the Suppression of Enemy Air Defences (SEAD) function. Without the F-22 to protect it, the F-35 is less capable as a fighter than fourth generation fighters, such as the F-16, that have been flying for 40 years.

one on the centreline. This made the aircraft wider, draggy, slower and less manoeuvrable. In short, the F-35 can't turn, can't climb, can't run.

In fact, it isn't a fighter aircraft in the first place. Australia might think it is buying a fighter than can hold its own against the Su-30, J-11, T-50, J20, J31 and others but it is really a light bomber. It was designed as such from the get-go. The recently retired head of Air Combat Command for the US Air Force, General Mike Hostage, has been quoted as saying, "The F-35 is geared to go out and take down the surface targets." The original requirement that evolved into the F-35 was Battlefield Interdiction and Close Air Support with the intent being to deal with lightly defended ground targets after the F-22 knocks out the really dangerous air defences. That assumes that a lot of F-22s are available. They aren't because production was halted at 187 in 2012. Two have crashed leaving 185 in service.

In the air combat role, Hostage says that it takes eight F-35s to do what two F-22s can handle. He has said further of the F-35: "Because it can't turn and run away, it's got to have support from other F-35s. So I'm going to need eight F-35s to go after a target that I might only need two Raptors to go after. But the F-35s can be equally or more effective against that site than the Raptor can because of the synergistic effects of the platform." He has also been quoted as saying that an F-35 pilot who engages in a dogfight has made a mistake. Further from General Hostage, "If I do not keep that F-22 fleet viable, the F-35 fleet frankly will be irrelevant. The F-35 is not built as an air superiority platform. It needs the F-22. Because I have such a pitifully tiny fleet (of F-22s), I've got to ensure I will have every single one of those F-22s as capable as it possibly can be." It therefore follows that Australia, not having F-22s, has an irrelevant air combat capability in the F-35.

The F-35's primary role in ground attack is confirmed by its weapons bays which each have room for a 2,000 lb bomb and one air-to-air missile. It could carry more bombs and missiles on its wings at the cost of stealth. At the same time, stealth against radar isn't the be all and end all of aerial combat. The F-35 can be spotted by low frequency radar a couple of hundred kilometers away. Infrared detection can also work at a considerable distance under the right atmospheric conditions. For example, the latest have infrared-scan-and-track system for the Sukhoi Su-27, the OLS-35, will detect, track and engage the F-35 at about 70 kilometres.

Due to severe transonic buffeting, wing roll-off and low acceleration, the F-35 is essentially a subsonic aircraft in both air intercept and ground attack missions. It cannot achieve supercruise as typically defined (sustaining speeds above Mach 1 without afterburner). All F-35 variants also have a very high infrared signature due to the hugely powerful engine required to push its brutal shape through the air, un-aerodynamic airframe and lack of infrared signature reduction measures. The problem is made worse by the fact that the F-35 has very limited rearward visibility, compounded by a large helmet that restricts head-turning. This will make surprise bounces from the rear quadrant a certainty. The only advantage that the F-35 has over the F-22 is presence of infrared-search-and-track, but the system in question is optimised for the ground attack, and so has limited air-to-air performance (limited ability to detect targets at a higher altitude than the F-35, limited range and resolution).

The F-35A has combat weight of 18.3 tonnes, a wing loading of 428 kg/m2, thrust-to-weight ratio of 1.07 and span loading of 1.75

tonnes/m. Wing sweep is 34°, and the engine has a power-to-frontal area ratio of 17.9 N/cm2. As a result, the F-35 has very bad instantaneous and sustained turn rates (50 per cent of the F-22's sustained turn rate, or ~14° per second) as well as bad acceleration, while its weight harms the transient performance. The F-35's inefficient aerodynamics and powerplant also limits combat endurance despite an excellent fuel fraction of 0.38.

The F-35 uses the GAU-22/A gun as well as AIM-9 Sidewinder within-visual-range missiles and AIM-120 beyond-visual-range missiles, though only the latter will be typically carried. The GAU-22/A needs 0.4 seconds to spin up to full rate of fire and the gun doors require 0.5 seconds to open. In the first second it will fire 16 projectiles weighing 2.94 kg. Again, usage of radar-guided missiles does not allow it to surprise the enemy at beyond-visual-range, and unlike the F-22, it can only carry an infrared missile at wingtip stations, thus negating its radar stealth.

The F-35 is far worse when it comes to damage tolerance than any other modern fighter, with massive quantities of fuel surrounding the engine inlet. This fuel will be at an elevated temperature during flight, and especially during combat, as it is used as a heat sink. The same fuel is used in aircraft's hydraulic system. A hit from a 30 mm high explosive-incendiary round, as used by most Russian and Chinese fighters as well as Dassault Rafale, is almost certain to ignite the fuel and catastrophically destroy the aircraft. The engine is likely to ignite it even if the hit itself doesn't.

The F-35 has one system, still in development, that has considerable potential if it ends up working as promised. This is the Distributed Aperture System that allows the pilot to see all around the aircraft in every direction. The view is displayed inside the pilot's visor using data from cameras around the aircraft. Each helmet is made to fit the head of the pilot who will use it, at a cost of US$600,000 per helmet. The system allows the pilot to see through the floor of the aircraft and see the ground underneath. It also analyses all the other information coming in from the radar and the infrared cameras also around the aircraft and presents it on the field of view, along with similar data from other F-35s our pilot is flying with. The system determines what each threat is, ranks them in priority and recommends what countermeasure should be used. The F-35 can fire air-to-air missiles against aircraft flying behind it that the pilot cannot see. This is claimed in theory but will not work in practice. The beyond-visual-range AIM-120 missile that the F-35 will be carrying does not have the ability to do a 180° reversal and it needs mid-course guidance from the radar which is facing the other direction. Firing a missile "over-the-shoulder" consumes enormous energy and greatly reduces range.

Flying as a pack of at least eight, F-35s in theory should be able to provide mutual fire support. The F-35 could also serve as a sort of mini-AWACS directing 4th generation fighters such as the F-18 onto targets. That said, other aircraft, already in service, do the same thing. All the Sukhois and the Swedish Gripen have intra-flight data sharing and are truly mini-AWACs. Gripens are optimised for 'cloud shooting' so one aircraft targets and another passive aircraft (not emitting a radar signal) shoots.

The F-35 is a complicated aircraft though and may prove to have been just too ambitious. Its software includes over 30 million lines of code which is six times more than that of the F-18E/F Super Hornet. There are plenty of bugs in the software and the aircraft's other systems which will take years to work through.

One of the more important bugs is the helmet vision system which isn't as seamless as it needs to be and produces too many false alarms. And if the helmet isn't fixed, it definitely won't be a fighter because the aircraft's bulkhead behind the pilot continues at the same height as the canopy. The pilot wouldn't be able to see what's behind him if the helmet is not working. He also wouldn't be able to see below him because the aircraft is too wide. Most fighters have the pilot sitting up where he can see as much as possible. The F-35 pilot's head is down in the fuselage, like a bomber.

Australia's recent acquisition of a couple of large amphibious assault ships, the *Canberra* class of 20,000 tonnes, briefly raised the notion that they could be equipped with the short-takeoff-vertical-landing

(STOVL) version of the F-35 to provide combat air patrol and close air support (CAS). With respect to CAS, the A-10 aircraft, dedicated to that role, carries 1,350 rounds of 30 mm ammunition. By comparison, due to the compromises necessary to get the STOVL version to fly, the gun of the F-35 STOVL version is carried externally in a pod. It will hold 180 rounds of 25 mm ammunition weighing about 90 kg. The gun could burn through its ammunition load in three seconds. The STOVL F-35 is an expensive way of carrying 90 kg of ordnance into battle. It carries two 1,000 lb bombs instead of the 2,000 lb bombs on the air force version, once again due to weight limitations. There are likely to be far more cost efficient ways of providing fire support to the troops landed from our LHDs. While we are the subject of the F-35's gun, the software for the aircraft to be able to fire it won't be ready until 2017, and possibly 2019 by some reports. The software to enable the STOVL F-35 to drop the Small Diameter Bomb II, short enough to fit the bomb bay, won't be uploaded until 2022. Enthusiasts for using the F-35B on Australia's LHDs should be aware of its cost which is US$251 million each. The cost of the US Navy variant, the F-35C, is now US$337 million. This variant needs larger wings so its sink rate on approaching an aircraft carrier to land on is low enough and it can land at a lower speed.

A good summary of the current status of the F-35's bugs and shortcomings is provided by the US-based Project on Government Oversight (POGO), from a Department of Defense report.[1] The US Defense procurement system requires that weapons development programs remain on schedule or they become in danger of being scrapped. The F-35 is well behind schedule but production has begun before testing has been completed. POGO's analysis shows that Lockheed Martin, the aircraft's developer, has been cooking the test results to meet project milestones. The effect of that will be an expensive retrofitting of completed aircraft estimated at US$60 billion. For Australia, that might mean a further $30 million per aircraft delivered. That extra $30 million per aircraft is equivalent to half the cost of a Gripen.

There is an incident in the POGO report which suggests that the

F-35 might be fatally flawed because of the compromises made to get the thing to fly in the first place. In June 2014, there was an engine fire in an F-35 that was taxiing which resulted in the aircraft being lost. The aircraft that blew up was damaged, three weeks earlier, during two seconds of flight when the test pilot, operating well within the safety envelope of the aircraft's abilities in a ridge roll maneuver, put G forces, yaw and roll stresses on the aircraft all at the same time. The F-35's engine is said to have the problem of being too flexible. That may be because the airframe is too light, in which case this is a problem that is baked in the cake. There are severe flight restrictions as a result. If you put a fighter into a snap turn to (say) avoid a missile, the gyroscopic forces are huge. Both the engine and the aircraft have weight problems, and beefing up either or both compromises the already overweight aircraft. The practical outcome of that will be that the F-35 will be restricted in its manoeuverability by its software.

Another restriction is a limit of Mach 0.8-0.9 at low altitude because the F-35 cannot dissipate its heat. Its competitors are limited to about Mach 1.2, so if there is a low-altitude engagement, 'can't run' becomes a serious threat to its survival. In fact in battle simulations of the F-35 against the Su-35, 2.4 F-35s are lost for each Su-35 shot down. Pitting the Gripen against the Su-35 results in 1.6 of the Sukhois shot down for each Gripen lost. The loss exchange ratio of the Gripen against the F-35 is breathtaking. For each Gripen lost in a Gripen on F-35 exchange, 21 F-35s are shot down.

There is potentially a big positive outcome from all of this. Australia has a great need for an aircraft that can fly long distances straight and level without stressing the airframe – to fulfill the maritime strike role. We need an aircraft that can deliver anti-ship cruise missiles. The F-35 is already being prepared for that role with our Department of Defence and the Norwegian firm Kongsberg collaborating on fitting the latter's Joint Strike Missile to the aircraft internally and externally. This missile will have a range of 240 kilometres which means that the F-35 could launch it beyond the range of most ships' anti-aircraft missile systems. For the same role, the US is concentrating on its Long Range Anti-Ship

Missile (LRASM) which has a range of 930 kilometres. These could be launched from our F-18 Super Hornets as well as the F-35.

There is one view in the defence community that the F-35 program will die of embarrassment before the production of about 500 aircraft. This will leave a gaping hole in many countries' procurement schedules and there will be a mad scramble for supply from the European fighter makers. Australia started to buy Super Hornets when it became apparent that the F-35 schedule was slipping, first 24 and then another 12. Any F-35s that do turn up can be usefully applied to the maritime strike role.

An issue that affects all the international partners in the F-35 involves access to the computer software codes for the aircraft. The F-35 relies heavily on software for operation of radar, weapons, flight controls and also maintenance. The US military has stated that "no country involved in the development of the jets will have access to the software codes" and has indicated that all software upgrades will be done in the US. The US government acknowledges that Australia, Britain, Canada, Denmark, Italy, the Netherlands, Norway and Turkey have all expressed dissatisfaction with that unilateral US decision.

How will the F-35 go in actual combat? In the air-to-air role, the F-35 is woefully under-armed. It could carry more missiles on its wings at the expense of losing its stealth but otherwise it is limited to two beyond-visual-range missiles in its bomb bays. On encountering enemy aircraft, its best chance is to fire those two missiles at the earliest opportunity and then turn tail and run as fast as possible. Firing two beyond-visual-range missiles, each with a probability of a kill of 10 per cent, has a 19 per cent chance of downing one enemy aircraft. As General Hostage said, an F-35 that is in an aerial dogfight has made a mistake. They will be "clubbed like baby seals". In 2008, Major Richard Koch,[3] then chief of the US Air Force's advanced air dominance branch is reported to have said "I wake up in a cold sweat at the thought of the F-35 going in with only two air-dominance weapons."

The view that guns were redundant in aerial warfare following the development of air-to-air missiles first took hold in the 1960s. But missiles didn't perform as expected and most missiles missed. So the aircraft involved proceeded to the merge in which guns and the pure fighter attributes of manoeuvrability and turn rate were critical to survival. That remains just as true today.

If the F-35 is such a dog that is going to get a lot of RAAF pilots killed, why hasn't the program been terminated yet? The answer to that question is in the way the US defence procurement system is structured. Lockheed Martin, the plane's manufacturer, has spread production of components around 45 of the 50 states of the United States. The program was set up to be politically hard to kill, no matter what its shortcomings. We are going to pay the price of that.

Super Hornet

As the F-35 schedule started slipping a decade ago, the RAAF started acquiring Super Hornets to ensure against a capability gap emerging. The Super Hornet is based on the design of the F/A-18 Hornet but is much larger. In the 1990s, the US Navy needed a replacement for the swing-wing F-14 Tomcat. US legislation requires a competitive fly-off in choosing a new aircraft. The US Navy though couldn't afford to pay for the development costs of two aircraft so it got around that by promoting the Super Hornet as a design upgrade from the legacy Hornet

Figure 16: F/A-18E Super Hornet
Current RAAF policy is to continue to buy more Super Hornets as the F-35 schedule continues to slip. That is a mistake on two levels – the Super Hornet is outclassed by the Sukhoi Su-27 and its derivatives and there is a much more capable aircraft available that is also much cheaper, the Saab Gripen E.

rather than as a new aircraft, which it is. It shares only limited structural commonality with the F/A-18. While the Super Hornet forward fuselage is derived from the legacy Hornet design, the wings, centre and after fuselage, tail surfaces and engines are entirely new. It weighs 30 per cent more and has a 36 per cent greater internal fuel load.

Unfortunately for Australia, the Super Hornet is outclassed by the Sukhoi Su-27 Flanker and its later variants that China and some other countries in our region have in their air forces. The Flanker outperforms the Super Hornet in aerodynamic performance which will give it an advantage in within-visual-range combat. The Flanker's high supercruise speed combined with its large weapons load, enabling it to fling off four-missile-volleys at a time, also gives it the advantage in beyond-visual-range combat.

RAAF policy is to continue to buy more Super Hornets as the F-35 schedule continues to slip. That is a mistake. There is a more effective and cheaper alternative now available which we should be buying instead no matter what happens to the F-35. This is the Saab Gripen.

Saab Gripen

Saab was formed to produce aircraft in 1937. Their first jet fighter, the Tunnan, flew in 1950 with the first delta wing jet fighter, the Draken, following in 1955. Canards were added with the Viggen which first flew in 1967. The Viggen weighed 9.5 tonnes. It was followed by the Gripen, with an empty weight of 6.8 tonnes, introduced into service in 1997. This grew to 8.0 tonnes with the latest variant, the Gripen E. As with its predecessors, the Gripen was originally designed for flexible deployment with a small logistical footprint. This was due to the Swedish Air Force's policy during the cold war to operate out of a number of dispersed bases across the country. It was considered vital to keep staff resources, support systems and spares to a minimum. As a result of this, Gripen was designed to operate from runways only 800 metres long by 16 metres wide. This means it can land on a regular highway, which further improves its logistical flexibility.

The maximum combat radius for Gripen E on an air-to-surface configuration is approximately 1,500 kilometres. This is defined as flying to a target, releasing air-to-surface weapons and then returning to home base. By comparison, the F-35's range on the same basis is 1,100 kilometres. The actual combat radius depends on the configuration of the aircraft's external stores and the availability of reserve fuel tanks. In a typical air-to-air configuration for example, the Gripen E can patrol for over two hours. The one-way ferry range of the Gripen E is 4,000 kilometres. The Gripen is also designed for ease of maintenance with a turnaround time of 10 minutes in the air-to-air configuration. It can be re-armed and refuelled with the engine running. Maximum speed of Mach 2 is 0.4 Mach higher than that of the F-35. The Gripen uses the same engine as the Super Hornet.

With respect to air superiority, the Gripen has performed very well against the F-22 in Red Flag exercises in Alaska, possibly besting it. The Gripens had the F-22s in visual range before the F-22s were aware of their presence. Australia would lose nothing in terms of air-to-air performance by opting for the Gripen. Then there is the force-multiplier effect of the cost advantages, three of them.

Firstly, the Gripen E has a list price of US$70 million per copy though that comes with a generous spares package. On a like-versus-like basis, the price may be of the order of US$60 million. By comparison, one F-35A costs US$148 million and there will be a further US$30 million to come on top of that for doing rework to delivered aircraft. At a most likely arrived cost of US$180 million for the F-35A variant that Australia is getting, we could buy three Gripens.

Also the Gripen can mount sorties at twice the rate of the F-35 because of its lower maintenance requirement. So for every F-35 sortie we could mount six Gripen sorties for the same capital outlay. We would not be sacrificing quality for quantity in doing this. Remember that the Gripen can shoot down 21 F-35s for each Gripen lost to an F-35. Adding that factor in and the Gripen becomes 126 times as cost-effective as the F-35.

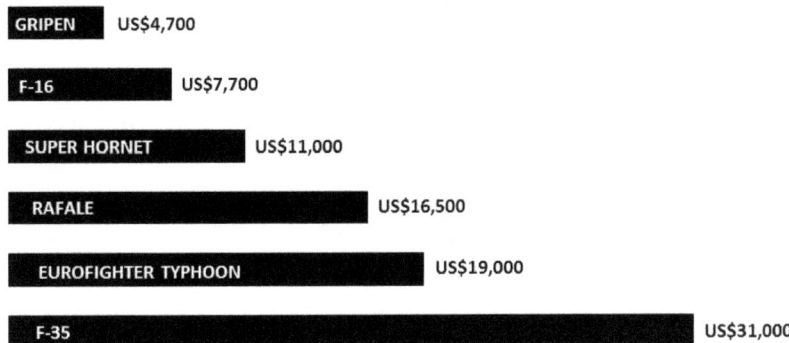

Then there is the flying cost per hour difference as shown by the graphic above.

Each hour of flight in an F-35 costs more than six times what it does in a Gripen. Australia won't be able to afford to fly its F-35s much at all, and certainly not enough to make our pilots fully proficient in flying them. We could afford lots of flying time for pilots in the Gripen. This would make such Gripen pilots far more effective in combat. That is difficult to quantify but the higher level of training could easily make pilots 40 per cent more effective. Adding this factor in means that the Gripen becomes 176 times as cost-effective as the F-35. And the Gripen will make far fewer widows than the kamikaze flights of the F-35 will, because flying the F-35 against the Sukhoi Su-27 and its later variants will be mostly a one way trip. On top of all that, adopting the Gripen means that Australia would be secure by achieving air superiority at least out to the Indonesian archipelago. If we rely upon the F-35, we won't have an air force at all because every other type of aircraft we have will be shot down as well.

The case for the Gripen over the F-35 is overwhelming. Fortunately, while Australia has announced an intention to acquire a total of 72 F-35s, so far we are only contractually bound to buying two of them. So we could back out of the worst of this horrendous mistake if we had the will and the wit to do so. If we persist with the F-35, most likely we will only get a fraction of that number of 72 aircraft before production is halted due to the F-35 being too deficient to continue with. Any F-35s that arrive in Australia can be mated with anti-ship cruise missiles and external fuel tanks and applied to the maritime strike role. They would be safe out at sea where they won't encounter enemy aircraft. They will be an expensive way of delivering cruise missiles because the Gripen could do that far more cheaply as well. The pragmatic way of dealing with the F-35s would be to park them up, write off the capital cost and buy an equal number of Gripens. It would still be cheaper than trying to fly the F-35s and we would have a far more effective force.

How many Gripens should we buy? When Australia had half its current population and GDP per capita was a lot lower in real terms also, we had 113 Mirage IIIs in service. By that metric we could have at least 200 Gripens. The F-35s we are buying will cost us more than $250 million each. The 72 on order in total will cost us more than $18 billion. That same lump of money would buy us 222 Gripens which is much the same as the number derived by maintaining force equivalence per capita. Buying that number would mean that they could be built in Australia as the Mirage IIIs were. A number of air forces around the world will be left short when the F-35 program collapses. There is a good chance we will end up building for export as well.

You could be forgiven for being incredulous about what you have just read about the F-35. Surely such an expensive and important programme would have had a lot of thought go into it. But the Department of Defence has had plenty of inappropriate acquisitions and failed projects in its recent history. The *Collins* class submarine is the most prominent of these but perhaps the worst in terms of outlay versus result was the $1.4 billion spent trying to upgrade Super Seasprite helicopters for the Navy before the project was cancelled in 2009. The F-35 is simply the Super Seasprite project writ large – costs blowout and the finished product falls well short of what is required. It was conceived at about the same time and signed off by the same people. Also, the Department of Defence believes in global warming and actually has staff assigned to worry about that. That indicates that a lot more intellectual rigour could be applied to the Department's efforts. Anyone who believes in global warming has enormous capacity for self-delusion as well as a lack of rigour.

Bomber aircraft

Australia did have a strategic bomber force of 24 F-111s up to 2011. They were taken out of service on the basis that they were becoming maintenance-intensive with age. That is most likely just an excuse for the fact that the F-35 programme is eating the budgets of other aircraft. By comparison, the US is extending the life of its B-52 bombers to 2040, by which time they will be 80 years old and will have been used by three generations of servicemen. The F/A-18, the Super Hornet, the F-35 and the Gripen can all drop bombs and fire cruise missiles. The F-111 had longer legs though and was a more cost-effective platform for delivering bombs, but it is no more.

What actually happened to the F-111s points to deliberate vandalism to remove competition for the Super Hornet and the F-35. In the fatigue testing of the wing element that justified the decision to scrap the F-111, the test article was overloaded and it consequently broke. Australia had purchased massive supplies of spares for practically nothing after the US Air Force withdrew the F-111 from service. So, even if we had to change wings, there were plenty in store. In addition, the F-111 wings could be accessed by reskinning. The only life limitation was the wing carry-through box at something like 80,000 hours of flight. The US Air Force had done the engineering work to replace the TF-30 engines of the original F-111 with the F100 engine which powers the F-15 and F-16, increasing the range by 60 per cent.

When the RAAF's Mirage IIIs were retired, they were flown off to desert storage in South Australia. Instead of doing that, the RAAF cut up the F-111s and buried them in landfill in the Ipswich suburb of Swanbank. It seems the RAAF didn't want to have the possibility that the F-111s might rise from the dead. On top of all that, the US Air Force at the same time pulled all its F-111s out of storage in the "Boneyard" and scrapped them, and they never do that. The Boneyard is the Aircraft Maintenance and Regeneration Centre at Davis-Monthan Air Force Base in Tucson, Arizona, where all US military aircraft are sent in retirement. Aircraft can be resurrected after a long time in the Boneyard. For example, in 2009 a Canberra bomber that had been built under licence in the United States was brought back to serve in Afghanistan after 40 years in the Boneyard. Scrapping the F-111s got rid of a couple of billion dollars' worth of capability so that we would have to buy new aircraft.

All these incidents suggest a very close relationship between the RAAF and the US Air Force which is of course a very good thing at one level. But it seems that the RAAF has become too close and has been morally corrupted to serve the interests of Lockheed Martin – why otherwise would have the RAAF gone to such lengths, contrary to established practice, to literally bury the F-111? From the foregoing it is easy enough to guess that promotion within the RAAF would require mindless adherence to the belief that the F-35 is the answer to a maiden's prayer. This institutional blindness will cost Australia dearly. The Federal Government would not be getting impartial advice on the relative merits of weapons systems.

Australia does have a need for a platform that can cost-effectively deliver bombs and cruise missile at a considerable distance from our shores. If we leave interdiction of enemy surface forces until they are close to Australia's coastline then that will result in a lot of dead Australians from attacks by those enemy forces. That said, if survival of fighter aircraft is problematic in the modern age then that is doubly so for bomber-sized aircraft. Even if they were stealthy, they could still be detected by UHF and VHF radars from several hundred kilometres away. The most advanced Russian surface-to-air missile system, the S-400 Triumf, has a range of 400 kilometres. Ramjet-powered beyond-visual-range air-to-air missiles can reach Mach 4 and have a range of more than 200 kilometres. Fighters directed by over-the-horizon radar and firing those sort of missiles will start picking off bombers a long way out from their targets.

The United States has plans to develop a long range strike bomber which it hopes will have a price per copy of about US$500 million. A bomber that is still thought to be capable of penetrating modern air defences is the B-2 Spirit, a batwing aircraft without vertical tail surfaces. Each B-2 cost US$2.1 billion to build and has an operating cost of US$135,000 per hour of flight. It also needs to be kept in an air-conditioned hanger. The United States could only afford to build 21 of them, of which 20 remain.

There is an effective alternative which was first thought of back in the 1980s at a time when the cost of building bombers blew out. This is to convert commercial airliners to dropping cruise missiles. Boeing proposed to convert Boeing 747s to carry 72 cruise missiles on rotary racks. The cruise missiles would be ejected from a port on the rear starboard side of the aircraft. More recently, the Boeing 737 has been used as the airframe for the P-8 maritime surveillance aircraft. The P-8 can drop torpedoes from a bomb bay in the forward fuselage.

Boeing's list price for a 737-800 is US$93 million. Second hand 737s with plenty of life still left in them can be had for under $10 million. Cutting a port in the ribbed airframe and reinforcing it might cost $2 million. Racking and handling systems for the cruise missiles to be carried might be another $2 million. Including added electronics, the cost of the converted 737 might be of the order of $30 million. It could carry 15 cruise missiles weighing 1.3 tonnes each, the weight of the current Tomahawk cruise missile. With a range of more than 5,000 kilometres return and a range of the cruise missile of 1,000 kilometres, a 737 so configured flying from RAAF Base Tindal in the Northern Territory could attacks targets as far away as Beijing. It is more important though to conduct saturation attacks against enemy naval forces. A flight of four 737 bombers could launch 60 cruise missiles against an enemy surface action group which would overwhelm the defences with the leakers getting through to do the damage.

Just as civilian 737s fly night and day with a change of crew, 737 bombers could be conducting two sorties a day against targets in the South China Sea. With a cruising speed of 828 kilometres per hour, our 737 bombers could be rapidly vectored against targets of opportunity. The cruise missile's range of in excess of 1,000 kilometres means that the 737 bomber need not have to operate anywhere where it might be in danger from enemy aircraft or surface-to-air missiles. That said, China has one working high-frequency-skywave radar and is rapidly commissioning more to form a network. The combination of that skywave radar network and the J-20 fighter would make survival within 1,000 kilometres of the Chinese coast problematic. The solution to that would be to increase the range of the cruise missiles carried. Cruise missiles have a fuel economy of three kilometres of flight per litre of fuel, so for a given weight, range is a trade-off between fuel capacity and warhead size.

Naval vessels exist to sink the enemy's ships but a high proportion of their systems are devoted to protecting the vessel from enemy aircraft and submarines, reducing the amount of ordnance that can be fired at enemy ships. That ordnance is missiles and cruise missiles fired from vertical launch cells. Take the example of Australia's new Hobart class air warfare destroyers, each of which will have 48 of the Mark 41 vertical launch cells and eight Harpoon missiles in canisters. If half of the vertical launch cells were devoted to cruise missiles, then the capital cost of taking cruise missiles contained in those launch cells into battle works out to about $120 million per cruise missile. Once the vertical launch cell is expended, the ship has to go back to port to have it reloaded by a crane on the dock. By comparison, the capital cost of carrying one cruise missile into battle by 737 bomber would be about $2 million. And the 737 bomber could be reloaded twice a day and will hold a far greater area at risk than the destroyer.

The first cruise missile was the V-1 during the Second World War. English engineering analysis at the time suggested that the cost of making a V-1 was equivalent to that of making a car. The same relationship should hold true today as cruise missiles have very few moving parts. Other than the small turbofan, they are mostly a fuel tank and a warhead, yet they cost ten to twenty times what a modern car costs. That may be mostly because they are built to the high engineering standards of the aerospace industry which now has very low failure rates. Given the numbers of cruise missiles we should be launching, we could possibly get the cost of cruise missiles down considerably by adopting a slightly higher failure rate.

Transport aircraft

The RAAF has 30 transport aircraft in service and on order. These are eight C-17 Globemaster aircraft capable of carrying 77 tonnes, 12 C-130J Hercules capable of carrying 19 tonnes and ten C-295 Spartan capable of carrying 11.5 tonnes. The virtue of the C-17 Globemaster is that it can carry whole tanks at an operating cost of US$21,000 per hour. The C-130 Hercules can carry many types of vehicles and can operate from short airstrips. The C-27J Spartan from Alenia in Italy has one competitor in its class – the C-295 made in Barcelona by Airbus. The latter is much more cost effective but the C-27J is slightly larger and can carry vehicles that the C-295 can't. It seems that Australia bought the C-27J instead of the C-295 because the former was easier to load. The ten C-27J that Australia has ordered, with two of those delivered to date, are replacing 29 Caribou transport aircraft that first saw service with the RAAF in 1964 and were retired in 2009.

The characteristics of the RAAF's transport aircraft in service, and the ones we should buy, are tabled following:

		RAAF Nos.	Weight tonnes	Cost US$m	Payload tonnes	One Way Range km.	Payload/ Weight	Cost/ Tonne US$m
C-17	Globemaster	8	282	$250	77	4,482	0.27	0.89
C-130J	Hercules	12	34	$100	19	5,250	0.56	2.94
C-27J	Spartan	10	17	$60	11.5	1,850	0.68	3.53
C-295	Persuader		11	$28	9.2	1,300	0.84	2.55
PZL M28	Skytruck		4	$4	2.3	1,500	0.58	1.00

The problem is that we have only 30 transport aircraft for the whole of the vastness of northern Australia. Of those 30 aircraft, only 22 aircraft that can take off from short and rough airstrips. Come a war and there will be frenetic activity across Australia's north and no means of repositioning small groups quickly. We need a much larger airlift capacity, especially smaller transports that can shift a section of troops or do resupply runs. In addition to the existing force structure, Australia's vastness could easily soak up 30 of the C-295.

The C-295 is half the cost of the C-27J but its payload is only 20 per cent less. It also has half the fuel consumption per hour of flight and thus the C-295's cost per hour of flight is US$1,750 compared to US$2,542 for the C-27J. So the C-295 costs half as much to buy, and about one-third less to operate, that the C-27J. There is a role for such an aircraft in Australian service because we simply have a fraction of the numbers of transport aircraft we will need in a conflict. Our C-27Js were bought from US stock through the US Foreign Military Sales system instead of directly from the manufacturer and somehow ending up paying about 40 per cent more than we should have, so an indifferent purchase was compounded by a foolish way of buying them.

We also don't have the capability to move small groups of troops around by air or resupply outposts with small loads of supplies. Helicopters would be an expensive way of providing this support and we simply don't have enough of them either. The US Special Operations Command can buy whatever it deems appropriate and is mandated to go outside the normal military procurement process to do so. So, to move small numbers of troops around, Special Operations Command chose the PZL M28 Skytruck and bought ten of them. This design started life as the Antonov An-28 which first flew in 1969 in Ukraine.

Figure 17: C-17 Globemaster
The Globemaster is a 282 tonne aircraft that can carry a main battle tank.

The Skytruck is now produced in Poland by PZL Mielec which was acquired by the US aircraft company Sikorsky in 2007. Sikorsky has recently been acquired by Lockheed Martin. The Skytruck is a twin-engine, high-wing strutted monoplane of all-metal structure, with twin vertical fins and a robust tricycle fixed landing gear, featuring a steerable nose wheel to provide for operation from short, unprepared runways where hot or high altitude conditions may exist. The Skytruck's cruising speed is 270 kilometres per hour. Australia needs at least 50 of them, which could be procured for the cost of one F-35. We could easily do with two F-35s' worth.

The RAAF's transport fleet also includes two Boeing 737s and one Bombardier Challenger business jet in its VIP fleet and nine Beechcraft Super King Air for personnel transport and navigator training.

Airborne warning and control

The concept of using radar-equipped aircraft to coordinate other aircraft was first used late in the Second World War in the attacks on Japan. The heyday of airborne warning and control (AWAC) aircraft started in the 1970s with the development of the Boeing E-3 Sentry based on the Boeing 707 airframe. It is distinguished by its large radome on struts above the aircraft. This revolves at six revolutions per minute. The tactical use of AWACs evolved to standing back a couple of hundred kilometres from aerial combat and using its powerful radar to determine what was happening on the aerial battlefield. An AWAC can detect aircraft 400 kilometres away. By those same radar waves it is also advertising its presence at least 800 kilometres away. Nevertheless AWACs have been safe and none have been shot down in combat yet.

The Soviet Union and its arms customers resented the success of AWACs and begun to make long range missiles specifically to shoot them down. The Chinese J-20 stealth fighter is a "bomb truck" that also appears designed to shoot down AWAC aircraft and tankers in the first instance. A supercruise speed of Mach 1.6 at altitude is a ground speed of 1,900 kilometres/hour. Once it broke through the fighter

Figure 18: E-7A Wedgetail
The Wedgetail is an Airborne Warning and Control aircraft based on the Boeing 737 airframe which can detect fighter-sized aircraft 400 kilometres away.

screen that was engaging Chinese Su-27s and J-10s, a J-20 would cover 200 kilometres towards the AWACs in six minutes. At that point it would launch radar-homing missiles with a range of 200 kilometres on a ballistic trajectory at Mach 4. The AWAC crew would barely have time to get out via their escape chute. The tankers would also be shot down and at that point Allied aircraft would have to disengage, if they could, to avoid running out of fuel. The effect of all this would be to push out the Chinese-dominated air domain a couple of hundred kilometres eastward.

The RAAF asked for proposals for AWAC aircraft in 1996 and awarded a contract to Boeing in 2000 to develop the E-7A on the Boeing 737-700ER airframe. In Australian service it is known as the Wedgetail. It has also been adopted by the South Korean and Turkish air forces. The Wedgetail uses an electronically scanned array radar in a dorsal fin on top of the fuselage. The radar has a maximum range of 600 kilometres in detecting aircraft operating at altitude. When looking down at fighter-sized aircraft, the maximum range is of the order of 400 kilometres. It can see frigate-sized vessels on the ocean surface at a range of 240 kilometres. The Wedgetail can track 180 targets simultaneously and coordinate intercepts on 24 of these.

In some fighter types the role of increasing situational awareness has been taken on by the fighters themselves with the Su-27, Su-30MK series,

Su-35, T-50, F-35 and Gripen all having data links that enable sharing of radar and infrared target data within a group of aircraft. That is a good thing because AWACs may have to operate on the fartherest edge of their capability if they are to survive on the modern battlefield. Our Wedgetails will always be useful in detecting enemy shipping though and could be applied to that role if they are too vulnerable in their original role. We have six Wedgetails based at RAAF Base Williamtown which is 30 kilometres north of Newcastle.

Maritime patrol

Australia has 15 P-3 Orion aircraft for maritime patrol. The airframe is the Lockheed Elektra commercial airliner that first flew in 1957. It has a distinctive tail boom to house a magnetic anomaly detector for finding submarines. Apart from anti-submarine warfare, the P-3 Orion also performs maritime surveillance and can also engage surface targets. It has a bomb bay replacing a rear luggage compartment which can drop Mark 46 torpedoes, Stonefish naval mines and sonobuoys. Hardpoints on the wings can carry air-to-surface missiles.

Australia's P-3 Orion fleet peaked at 19. Currently 15 remain. They are being replaced by eight P-8 Poseidon aircraft with an option for four more. The P-8 Poseidon is based on the Boeing 737-800 airframe and is a much more capable aircraft. It is expected to enter service with the RAAF in 2017 with the last of the P-3 Orions being retired in 2019. Our maritime patrol aircraft are based at RAAF Base Edinburgh which is 25 kilometres north of Adelaide.

As a maritime nation, Australia has a big hole in our sea rescue capability. We would want to rescue downed aircrew and crews of warships sunk as quickly as possible. At the moment the only way we could do that is send patrol boats or frigates which would take days if they were available for the tasking in the first place. The best solution to that problem is to acquire four of the US-2 seaplane produced by ShinMaywa of Japan. This is a four-engined turboprop with a range of 4,700 kilometres and a cruising speed of 480 kilometres per hour.

Tanker aircraft

The RAAF has taken delivery of five Airbus A330 MRTT tanker aircraft. They are based on the Airbus A330 commercial airliner and in Australia service are designated the KC-30A. The KC-30A has an empty weight of 125 tonnes, a maximum take-off weight of 233 tonnes and can deliver 65 tonnes of fuel at a distance of 1,800 kilometres with two hours on station. It can refuel two aircraft simultaneously with booms that extend from wing nacelles. The tankers are based at the Amberley Airbase west of Brisbane.

Jindalee Over-The-Horizon Radar Network

Australia has at least one system we developed ourselves from first principles that has turned out to be quite good. This is the Jindalee Over-the-horizon Radar Network or JORN for short. Australia's efforts in over-the-horizon radar commenced in the 1970s as Project Geebung to evaluate the ionosphere for reflecting radar waves. That was followed in 1974 by an experimental radar transmitting array in the Harts Range east of Alice Springs and a receiving array near Alice Springs. This radar system was handed over to the RAAF as an operational system a decade later. In 1986 the decision was made to build two more radars, one at Longreach in western Queensland and one at Laverton in Western Australia. The radars are positioned so that their coverage begins near the Australia coastline. They are controlled by a centre at RAAF Base Edinburgh.

The transmitting arrays at Longreach and Laverton each have 28 elements with each driven by a 20 kW amplifier for a total power of 560 kW per station. This power is provided by diesel generator sets burning 150 litres an hour and thus the radars are not operated all the time because of the expense. As a consequence the Laverton station missed tracking the path of the Malaysian Airlines flight MH 370. Over-the-horizon radars operate on the Doppler principle in which an object can be detected if its motion towards or away from the radar is different from the movement of its surroundings. They are not good at detecting things moving tangentially across its field of view. The radar beam is

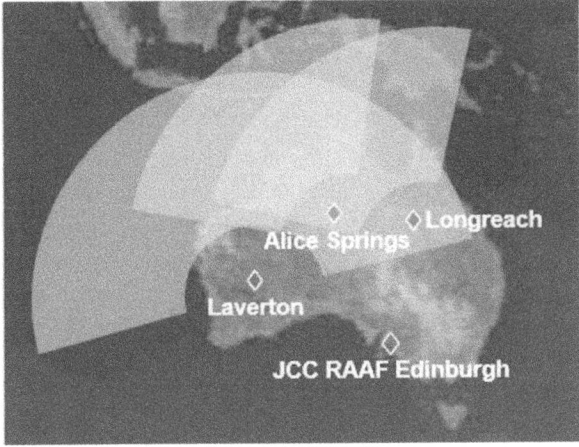

Figure 19: JORN Operating Coverage
This figure shows the location of Australia's Jindalee over the horizon radar stations and their control centre at RAAF Base Edinburgh. The transmitting and receiving bases are separated. For example the receiving base for the Laverton station is 81 kilometres due west of the transmitting station. As shown, the detection range is 3,000 kilometres but it is likely to be larger than that.

steered electronically to focus on one particular part of the total area of coverage at a time. This area being scanned is termed a tile. The radar's frequency can be changed to search for a particular type of target – Su 27s, Chinese bombers or frigates. The target resonates if it is of the same wavelength. The frequency can be changed from sweep to sweep, and hence search for and discriminate different types of target. Stealth only works for higher frequencies and thus the JORN system can detect stealth aircraft. Bigger aircraft like the F-35, F-22, B2, J-20 and T-50 have resonant frequencies in the 10-15 MHz range which is the sweet spot of high frequency over-the-horizon radar. The Gripen's resonant frequency is about 26 MHz which is rarely available due to ionospheric conditions.

In terms of what JORN can see, this is affected on a day-to-day basis by the state of the ionosphere. It is able to see Cessna-sized aircraft taking off in East Timor. In the 1980s it was able to see aircraft taking

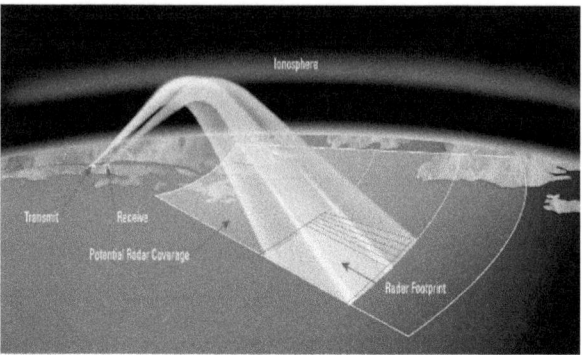

Figure 20: JORN Operating Principle
Instead of sweeping the entire sky like most radars, the JORN system focusses on one part of its coverage area at a time. The search frequency can be changed to look for aircraft of a particular size, or ships. It detects objects moving towards or away from the radar not objects moving perpendicular to its beam.

off from Singapore's Changi airport and the electronics have improved a lot since then. In 2007 it was able to detect a missile launch 5,500 kilometres away in China.

JORN's big test is coming up in the South China Sea where China is building three long airfields on Fiery Cross Reef, Subi Reef and Mischief Reef. At some stage China will declare an Air Defence Identification Zone over the South China Sea and those airfields will be used to enforce it. The airfields on artificial islands are another 1,000 kilometres beyond JORN's declared operational range of 3,000 kilometres but the system's electronics can be expected to be a lot more capable now. Whether or not JORN can consistently reach the South China Sea now, Australia should build at least two more JORN stations on our northernmost coastline, on the Broome Peninsular and Bathurst Island, to extend coverage over where the world's next major battle will be fought. The JORN system should be able to detect Chinese naval ships at that distance and provide targeting for Australian and Allied submarines. Hopefully JORN is now providing a feed for US Pacific Command in Hawaii and Guam and won't have to be patched in on the run after hostilities break out.

Satellite warfare

One Australian company, Electro Optic Systems, tracks every object in orbit of the Earth, at least down to the size of a baseball. The primary purpose of this is so that the company can sell the information to satellite operators so that they can avoid a collision with space debris. Space junk can also be moved by ablating it with a laser beam from a ground-based laser. The atoms boiling off one side of a piece of junk provide the motive force to move the piece to a more rapidly decaying orbit. Such lasers will also be useful for destroying enemy satellites.

Some Chinese weapons systems such as the DF-21 anti-ship ballistic missile rely upon reconnaissance satellites for their targeting information. The sooner these satellites are destroyed on the commencement of hostilities, the better. To that end, the United States is building its Space Fence satellite tracking facility on Kwajalein Atoll in the Marshall Islands southwest of Hawaii. This US$900 million facility will be completed in late 2017. The contract awarded to Lockheed Martin to build the S-band radar on Kwajalein Atoll also includes an option for the procurement of a second radar site in Western Australia. If all options are exercised, the contract is estimated to add up to more than US$1.5 billion over an eight-year period. The Western Australian facility will be built with the assistance of Electro Optic Systems.

These two locations, Kwajalein Atoll and Western Australia, suggest that Space Fence is primarily tasked with detecting and tracking Chinese satellite launches. From that it is only a small step to destroying them with ground-based lasers.

Up until the 1980s, long range missiles were equipped with inertial navigation systems and some also had terrain-matching radar. Then GPS became functional, enhancing missile accuracy. GPS satellites orbit at 20,000 kilometres above the Earth and do two orbits per day. A full constellation is 24 operational 95 per cent of the time. Russia has a similar system called GLONASS, now with a full constellation of 24 satellites. China's version of GPS is called BeiDou and is expected to be completed in 2020 with 35 satellites in orbit. The potential for the GPS satellites to be destroyed in orbit has pushed missile guidance systems to go back to including an inertial navigation system as a backup should the GPS signal fail.

Unmanned Aircraft

Australia will be buying between six and eight MQ-4C Triton unmanned aircraft with delivery starting in 2020. The Triton is a navalised version of the Global Hawk and is quite expensive at US$120 million for a 6.8 tonne aircraft. But it is also quite capable. The Triton can be airborne for over 24 hours at 55,000 feet at 610 kilometres per hour. It has a 360° X-band radar capable of surveying 5,200 square kilometres of ocean in a single sweep. This radar can identify a target in any weather condition and take high definition radar pictures and automatically identify the target using its own software. The Triton is semi-autonomous so that the operators can pick an area for the aircraft to fly in and set speed and altitude rather than having to operate the controls. One thing the Triton can do that the Global Hawk cannot do is rapidly descend to lower altitudes, built with a more robust lower fuselage to better withstand hail, birdstrike and lightning strike. It also has anti-icing systems on its wings. When at low altitude the Triton can use its laser designator and range finder. Presumably the Triton can designate targets for missiles fired from other aircraft.

Another thing the Triton can do is passively pick up and classify even faint radar signals. It is able to geo-locate these signals allowing mission planners to create an enemy "electronic order of battle" profile and keep the Triton and other aircraft out of the range of enemy radars and air defences. This will also be used for locating enemy warships for targeting.

Australian forces in Afghanistan leased several Israeli-made Heron unmanned surveillance aircraft. We now have two Herons based at RAAF Base Woomera to maintain continuity of aircrew experience in unmanned aircraft until the arrival of the Tritons. Strangely the Tritons will be located at RAAF Base Edinburgh, north of Adelaide. This basing is 2,300 kilometres, and four hours flying time, from the north-western Australia coast where they will be most useful.

4

Fuel Security

Introduction

One of the inspired tank commanders of World War 2, Heinz Guderian, made the observation that "The engine of the panzer is as much a weapon as the main gun." Well, if the engine is a weapon, so is the fuel tank and Australia's fuel tank is running on empty.

Back in the 1960s, there were plenty of fresh memories of World War II and rationing, particularly fuel rationing and shortages. Oil was understood to be a strategic mineral and that Australia was at risk because just about all of our oil needs were imported. So, at the time, the Federal government subsidised oil exploration – both seismic and drilling. And when the first large oilfields were found, in Bass Strait, they weren't commercial to develop because they couldn't compete with oil from the Middle East at US$4 per barrel so Australian motorists paid a premium for petrol to pass through a subsidy to Esso and BHP for their Bass Strait oil. As a result, we fared a lot better during the oil supply disruptions of the 1973 Yom Kippur War than we would have done otherwise.

Those Bass Strait discoveries were the start of 40 years of abundance during which Australia was largely self-sufficient in oil. That happy time ended in 2000 and our oil production has been falling away since. Similarly, we used to refine all the oil we consumed ourselves. Of the eight refineries that performed that task, the first closure was the Port Stanvac refinery in South Australia in 2003 and it has been recently followed by three more to leave just four – one in Queensland, none in the biggest market of New South Wales, two in Victoria and one in Western Australia.

Australia is now again in the parlous position of importing 90 per cent of its transport fuel requirements. Even if we were able to produce all the oil we needed, we wouldn't be able to refine it. At the same time, the geopolitical situation is deteriorating with ongoing wars in the Middle East and the prospect of a major war involving Asian countries that provide the bulk of our imported refined products.

There is a solution to Australia's liquid fuel security problem and it is a very good one. The solution is to use coal-to-liquids (CTL) technology to convert the low-grade portion of our substantial coal reserves into liquid fuels to keep farms, factories and food distribution systems operating. The CTL solution can fill the gap from our conventional oil production, is not constrained by biological inputs, can produce the entire range of fuels and chemicals needed with supply security enhanced by plants distributed around the country. It will obviate approximately $40 billion of annual fuel imports and provide a major boost to the economy and welfare of all Australians.

World Oil Supply Outlook

It has been said that the best economic forecast ever made was by a Shell Company geologist by the name of King Hubbert in 1956 when he predicted that US oil production from the lower 48 states would peak in 1970. Mr Hubbert used a methodology called the logistic decline plot based on the mathematics of extraction of a finite resource. US oil production duly peaked in 1970. The same methodology can be used to predict world oil production.

Reflecting the extraction of a finite resource, after some initial volatility, the plot settles down to a straight line. The intersection of that straight line with the x axis is the total volume of oil initially in the system. Half that figure is the year of peak production, in this case 2005. The oil price commenced rising faster from June 2004 as a result of the oil market moving from inherent oversupply to permanent tightness. This does not include traditionally higher cost liquid fuel sources such as oil sands, oil shale, shale oil and condensate.

Oil production outside the United States peaked a decade ago and is now in decline. Oil supply in the United States has risen in the last few years due to high oil prices and low cost finance. With drilling being curtailed due to the current low oil price, US oil production will now follow the rest of the world into long term decline as shown in Figure 21.

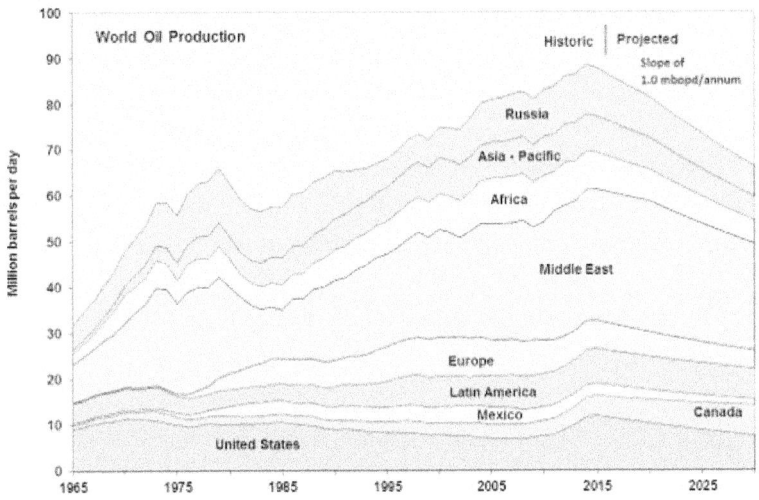

Figure 21: World Oil Production 1965 – 2030

Figure 21 shows Hubbert's peak in US oil production in 1970 and the European and Mexican production peaks. Production from the Middle East is expected to plateau over the next 20 years and thus oil supply will be further concentrated in the Middle East. The decline rate from here is expected to be about 1.5 million barrels per day per annum. One million barrels per day of oil production equates to 42 million tonnes per annum of LNG production in energy terms. Thus for LNG production to offset the decline in oil production, it would need to increase at about 60 million tonnes per annum.

Australia's Deteriorating Liquid Fuel Security

Following the development of the Bass Strait oilfields in the early 1970s followed by the oilfields off Western Australia, Australia was largely self-sufficient in oil production up to 2000, as shown in Figure 22. Those happy days are long past and Australia's oil production is now in long term decline to exhaustion. Our production of oil and condensate is forecast to decline from 147 million barrels in 2014 to 83 million barrels in 2030. In 2014, Australia imported 169 million barrels of oil and 132 million barrels of refined products. This equates to 465,000 barrels per day of oil and 361,000 barrels per day of refined products for a total of 826,000 barrels per day.

The increasing proportion of the world's oil supply coming from the Middle East increases the importance of the Straits of Hormuz chokepoint to Australian supply. The next major chokepoint is the Straits of Malacca, on the edge of the potential conflict zone in the South China Sea and East China Sea. In the event of a conflict, we are unlikely to be able to obtain any refined product supply from the Asian region.

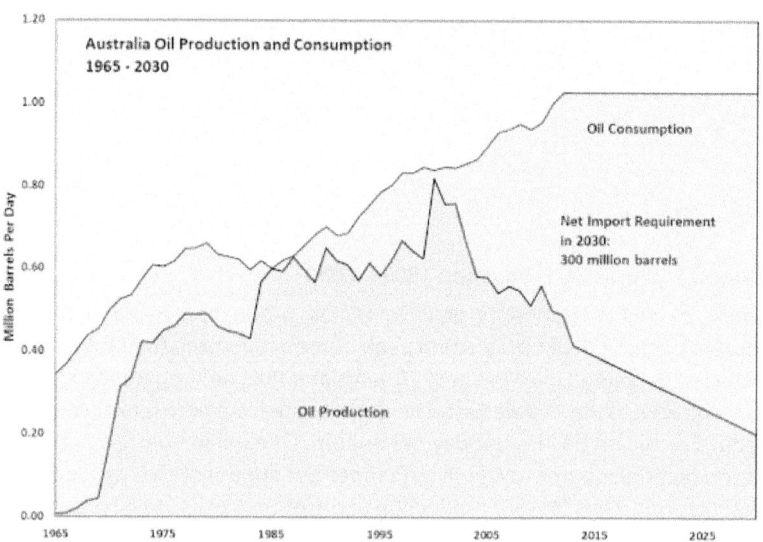

Figure 22: Australian Oil Production and Consumption 1965 – 2030

As well as being a national security problem and a potential cause of catastrophic economic disruption, Australia's crude oil and refined products imports are an enormous drag on the economy. In 2014, $41 billion left the country to pay for those imported fuels. At average weekly earnings, this equates to over half a million jobs which could be created if those funds stayed within the country.

Refinery Closures

Australia used to refine all the oil we consumed. Of the eight refineries that performed that task, the first closed in 2003 and it has been recently followed by three more, leaving four – one in Queensland, none in the biggest market of New South Wales, two in Victoria and one in Western Australia.

The consequences of those closures have begun in that fuel supply is being disrupted. In late January 2014, there was a shortage of jet fuel at Tullamarine airport because a ship cargo of fuel had been delayed. The refiners and marketers don't want to tie up a lot of capital in their business and tend to live a hand-to-mouth existence with just-in-time scheduling. Previously there had been a shortage in rural Western Australia due to a bad batch of imported diesel. In Victoria in 2012, there had been a shortage of diesel due to simultaneous outages of the two Victorian refineries.

Figure 23 (next page) shows what the evolution of Australia's refining industry looks like:

Fuel Quality

Energy density corresponds to the number of carbon atoms in molecules of each fuel type. Diesel has the highest energy density and is made of hydrocarbon molecules with between 10 and 20 carbon atoms in each molecule. Petrol has five to ten carbon atoms in its molecules. Jet fuel has 10 to 14 carbon atoms. LPG is made of propane and butane with three and four carbon atoms respectively. Ethanol has a low energy density due to the oxygen atom in its molecule.

Ethanol is also a poor fuel. With it having only 65 per cent of the energy density of petrol, pure petrol will take a vehicle 50 per cent farther than the amount of ethanol that takes up the same space in the fuel tank. Ethanol also has an enormous affinity for water, which it absorbs from the air above the fuel in the tank. Once the ethanol has become super-saturated with water, the ethanol-plus-water mix separates from the petrol at the bottom of the tank and can cause engine trouble. In the

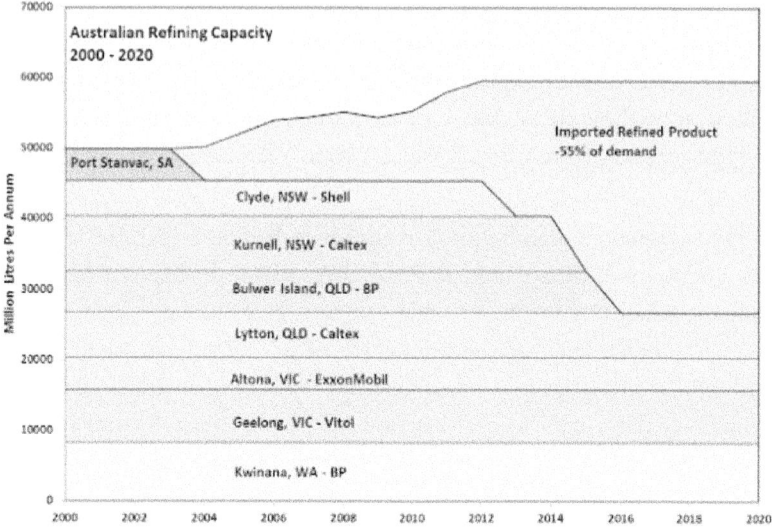

Figure 23: Australian Refining Capacity 2000 – 2020

Lack of refining capacity and lack of stocks means Australia will remain being highly dependent upon imports of refined products during any future war. Most of the refineries in Asia that supply most of our refined product imports are either in or on the edge of the future war zone in East Asia. This is an invidious position for Australia to have drifted into.

Australian environment, ethanol may take more energy to make than the energy it contains.

LNG has a relatively high energy density but requires storage at -160°C. In storage, methane is constantly boiling off and must be used or it will be wasted. Thus LNG is really only suitable for large vehicles that are in constant use. As such it will have limited market penetration. There is also no storage in the LNG production system with gas going from the wellhead through to the LNG plant in a continuous process. Once LNG is produced, it will require ongoing refrigeration in expensive tankage.

Another consideration with natural gas is that Australia's major gas fields are highly exposed off the coast of Western Australia. As such they are highly vulnerable to anti-ship cruise missiles, which today have ranges up to 1,500 km. It would be ill-advised to rely upon natural gas supply, and associated condensate and LPGs, in a conflict.

Compressed natural gas (CNG) has high penetrations of 70 per cent in the small vehicle fleets of Iran and Pakistan due to government mandates there. As a fuel it suffers from low fuel density and low transfer rates. Reticulated gas at low pressure has to be pumped up to the pressure in the vehicle's fuel tank. Due to the presence on the east coast of Australia of LNG plants that are short of their long term supply requirement, those plants will have the effect of making east coast gas trade at near parity with the oil price, which is now the driver of the LNG price in the Asian region. So reliance upon CNG will involve a lot of inconvenience, minimal storage ability and pricing near that of traditional oil products. Its mandated use in the Australian economy would result in a suppression of economy activity.

Despite high energy density for a battery, lithium ion batteries constitute 20 per cent of the weight of a commuter vehicle which means that 20 per cent of the energy is used for carting the battery around. In the absence of nuclear energy, the ultimate source of the electric power for electric vehicles is either natural gas burnt in gas turbines or coal burnt in power stations. Natural gas is already a suitable vehicle fuel almost straight from the wellhead. Burning it in a gas turbine leaves you with 40 per cent of the contained energy. In Australia, transmission losses take 10 per cent of what is left and charging losses take another 10%, leaving about 30 per cent of the original energy in the natural gas to drive the wheels. This compares to 50 per cent of the original energy if the natural gas had been used in a CNG vehicle. Thus electric vehicles waste 40 per cent of the available energy of the in the natural gas if it was used in vehicles directly. All the growth in the power market on the east coast of Australia has been gas turbines burning natural gas. As well as all that, CNG vehicles have a much longer range than electric vehicles. The range of CNG vehicles can also be extended using petrol tanks if they are dual-fuel.

With respect to burning coal in power stations to charge electric vehicles relative to turning coal into liquid fuels using the CTL route, the proportion of the original energy in the coal that gets to the wheels is much the same. The CTL route has far greater range and storage though. A tank of diesel will take a vehicle 50 times further than the same volume of lithium ion battery. Diesel can be stored indefinitely at any scale whereas electric power has to be used the instant it is created. Electric vehicles might be appropriate if the electric power source was from nuclear energy. Even then, using electric power to make liquid transport fuels may be a better proposition.

Mandated stockholdings

Australia is obliged to hold 90 days of imports as stocks, either as crude oil or refined products, as a signatory to the International Energy Agency treaty. Of the 29 countries that signed up to that treaty, Australia is the only one that is delinquent. We are currently delinquent to the extent of about 40 days of stocks and this is increasing as our domestic oil production declines. The IEA requirement is the minimum that we should hold under that treaty. Several Asian countries hold hundreds of days of stocks by comparison. If Australia were to build and to hold 50 days of refined product stocks at the current oil price and A$/US$0.73 exchange rate, the total cost including tankage at $50 per barrel of storage would be of the order of $18 billion.

According to an International Energy Agency report of 2013 entitled *Focus on Energy Security*, tank construction costs for above ground storage range from US$29 to US$37 per barrel, not including jetty construction costs. That report notes an Australian premium for construction costs of 20 per cent. Storing low flash point products such as petrol also adds a construction cost premium. Mandated stocks are a palliative and not a solution to transport fuel security. There won't be enough time during a conflict, while the stocks run down, to make other arrangements. The only viable solution to Australia's transport fuel security problem is CTL.

The CTL solution

CTL processes can convert coal to a range of fuels from methane through to the heavy distillates and beyond to waxes. There are three separate processes. Direct liquefaction of coal (Bergius process) and indirect liquefaction (Fischer-Tropsch process) were developed in Germany in the 1920s. The methanol to gasoline process (MTG) was developed by Mobil in the 1970s.

In direct liquefaction, hydrogen atoms are forced into the structure of coal molecules under pressure and temperature and in the presence of a catalyst. Wandoan coals in Queensland are likely to be good feedstocks for direct liquefaction because they start with a high hydrogen content. The hydrogen for the process is generated by steam shifting carbon monoxide to hydrogen and carbon dioxide.

Direct liquefaction plants tend to be complicated with a high capital cost. Only one modern plant has been built, by the Chinese coal mining company Shenhua in the Ordos Basin of Inner Mongolia. After that plant, Shenhua has gone on to build Fischer-Tropsch (FT) plants.

In an FT plant, coal is burnt in pure oxygen to produce a synthesis gas of carbon monoxide and hydrogen. The synthesis gas is then passed over a catalyst. Different catalysts will produced different suites of products. For example, an iron catalyst at a relatively low temperature will produce a high proportion of heavy waxes which in turn will need to be hydrotreated to produce lighter molecules in the jet fuel and diesel range.

There is one problem with the products from the FT process. The molecules in the diesel range that it produces have a specific gravity of about 0.76 while having a very good cetane number. The Australian specification for diesel is a specific gravity in the range of 0.82 to 0.85. Because of their zero sulphur, FT diesels will also require the addition of a lubricity agent though low sulphur diesels from crude oil also require that. The FT process also produces every molecule down to methane including LPGs and petrol.

In the MTG route, the synthesis gas is first made into methanol which is then passed through a zeolite catalyst. The pores in the catalyst limit the size of the molecules to having ten carbon atoms. Thus the MTG route cannot produce diesel and jet fuel. It does make a product though that is exactly on specification for petrol. The MTG route has a lower capital cost than the FT route and is simpler and cheaper to operate. Because the product from the catalyst is at the retail specification, no further refining steps are needed.

Thus in a CTL industry to make Australia self-sufficient in transport fuels, FT-based plants will provide the requirement for diesel and jet fuel and MTG plants, which could be more regionally distributed, would provide the balance of the petrol requirement.

CTL were built in Germany in the 1930s and in South Africa in the 1950s. The US built a large coal to synthetic natural gas (SNG) plant in North Dakota in 1984 in response to the second oil shock. There has been a new burst of activity in China in the last ten years, including 20 coal conversion projects in 2013. Chinese CTL production will rise to 1.1 million barrels per day by 2020, requiring 180 million tonnes of coal per annum. China also has 20 SNG projects in train which will require a further 200 million tonnes of coal per annum.

Rocks will burn in pure oxygen down to a carbon content of 10%. Thus low grade coals which aren't worth transporting can be used for CTL feedstock. High quality coals can produce up to 2.2 barrels per tonne. Lignite, which is 60 per cent water, can produce 0.6 barrels per tonne. The lignite resources of the Latrobe Valley would produce 60 billion barrels and this is the natural fate of that resource.

The capital cost of CTL processes is about $150,000 per daily barrel of capacity which equates to about $400 per annual barrel of capacity. The capital costs and jobs created by state are summarised in the table on the following page.

The total capital expenditure required is about $110 billion with the creation of about 30,000 jobs in the industry. The capital cost is twice what the three LNG plants built on Curtis Island at Gladstone cost but spread around the country instead of being concentrated on a strip a few kilometres long. But all the LNG plants built in Australia in the last few years show that the task of building the needed CTL plants is well within Australia's ability.

	FT Syncrude Barrels/Day	Petrol Barrels/Day	FT Capex $150,000/ daily barrel	Petrol Capex $120,000/ daily barrel	Direct Jobs	Indirect Jobs	Total Jobs
Queensland	195,000	35,000	29,250	4,200	3,067	6,133	9,200
New South Wales	201,000	70,000	30,150	8,400	3,613	7,227	10,840
Victoria	100,000	60,000	15,000	7,200	2,133	4,267	6,400
Tasmania		10,000		1,200	133	267	400
South Australia	47,000	14,000	7,050	1,680	813	1,627	2,440
Western Australia	50,000		7,500		667	1,333	2,000
Totals	593,000	189,000	88,950	22,680	10,427	20,853	31,280

Coal production to feed these plants would raise Australia's coal production by 50 per cent and create 27,000 jobs directly and 70,000 indirectly. That said, the inherent price volatility of the liquid fuels market means that support at all levels of government is needed to develop this nascent industry that is so vital to Australia's security and economy. The fuel excise levy can provide a mechanism. The Federal Government should make synthetic liquid fuels free of the fuel excise levy, aiding the economics of production by US$49 per barrel. Under this mechanism, the Federal Government does not have to pick winners. If a plant can't make product, it doesn't get assistance.

Climate is not a problem

While climate change is real in that the climate is always changing, and the greenhouse effect of carbon dioxide is also real, the effect at the current atmospheric concentration of carbon dioxide is minuscule. The greenhouse gases keep the planet 30°C warmer than it would otherwise be if they weren't in the atmosphere. So the average temperature of the planet's surface is 15°C instead of -15°C. Of that effect, 80 per cent is provided by water vapour, 10 per cent by carbon dioxide and methane, ozone and so on make up the remaining 10%. So the warming provided by carbon dioxide is three degrees. The pre-industrial level of carbon dioxide in the atmosphere was 286 parts per million.

Let's round that up to 300 parts per million to make the maths easier. You could be forgiven for thinking that if 300 parts per million produces three degrees of warming, the relationship is that every one hundred parts per million produces a degree of warming. We are adding 2 parts per million to the atmosphere each year which is 100 parts per million every 50 years and at that rate the world would heat up at a fair clip.

But the relationship isn't arithmetic, it is logarithmic. The University of Chicago has an online program called Modtran which allows you to put in an assumed atmospheric carbon dioxide content and it will tell you how much atmospheric heating that produces. It turns out that the first 20 parts per million produces half of the heating effect to date. The effect rapidly drops away as the carbon dioxide concentration increases.

By the time we get to the current level in the atmosphere of 400 parts per million, the heating effect is only 0.1°C per one hundred parts per million. At that rate, the temperature of the atmosphere might rise by 0.2°C every one hundred years. The relationship between atmospheric concentration and heating effect is shown in Figure 24.

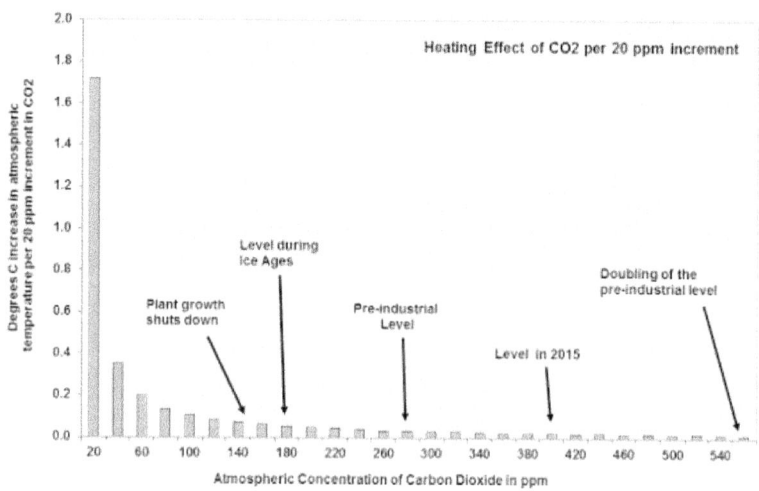

Figure 24: Heating Effect of CO2 per 20 ppm Increment

The total atmospheric heating from carbon dioxide to date is of the order of 0.1°C. By the time humanity has dug up all the rocks we can economically burn, and has burnt them, the total heating effect from carbon dioxide might be of the order of 0.4°C. This would take several centuries. A rise of this magnitude would be lost in the noise of the climate system. This agrees with observations which have not found any signature from carbon dioxide-related heating in the atmosphere.

The carbon dioxide level of the atmosphere is actually dangerously low, not dangerously high. During the glacial periods of our current ice age, the level fell to as low as 180 parts per million. Plant growth shuts down at 150 parts per million. Several times in the last three million years, life above sea level came within 30 parts per million of extinction due to a lack of carbon dioxide. The more humanity can increase the atmospheric concentration of carbon dioxide, the safer life on Earth will be.

Further to all that, belief in global warming from carbon dioxide requires a number of underlying assumptions. One of these is that the feedback loop of increased heating from carbon dioxide causes more water vapour to be held in the atmosphere which in turns causes more heating in a runaway effect. And that this feedback effect only starts from the pre-industrial level of carbon dioxide in the atmosphere – not a higher level or a lower level, but exactly at the pre-industrial level.

In the real world, there has been a temperature rise of 0.3°C in the last 35 years as measured by satellites. This is well short of what is predicted by global warming theory as practiced by the CSIRO, Bureau of Meteorology and others. There is also a far more plausible reason for the warming of the planet during the current Modern Warm Period which followed the ending of the Little Ice Age in 1900. The energy that keeps the Earth from looking like Pluto comes from the Sun and the level and make-up of that energy changes.

The Sun was more active in the second half of the 20th century than it had been in the previous 8,000 years. As shown by the geomagnetic Aa Index, the Sun started getting more active in the mid-19th century and the world's glaciers started retreating at about the same time. It is entirely rational to think that a more active Sun would result in a warmer Earth and this is borne out by empirical observation. To wit, the increased Antarctic sea ice cover observed during the satellite period.

Arctic sea ice extent retreated for the last 20 years of the 20th century. That is compatible with global warming from any cause. At the same time, Antarctic sea extent increased by an amount similar to the Arctic sea ice loss. This is not possible with global warming due to carbon dioxide. It also means that global warming due to carbon dioxide did not cause the bulk of the warming in the rest of the planet because carbon dioxide's effect was overwhelmed in Antarctica by some other force.

The increase in Antarctic sea ice extent is entirely consistent with increased global temperatures due to high solar activity as explained by Henrik Svensmark's theory. This theory holds that high solar activity produces a lower neutron flux in the lower troposphere from intergalactic cosmic radiation, in turn providing fewer nucleation sites for cloud droplet formation and thus less cloud cover. Sunnier skies over Antarctica in turn mean that more solar radiation is reflected by high-albedo snow and ice instead of being absorbed in the cloud cover. Thus Antarctica has cooled.

The rest of the world has enjoyed the best climatic conditions, and thus agricultural growing conditions, since the 13th century. But what the Sun gives it can also take away. Solar physicists have been warning for over a decade now that the Sun is entering a prolonged period of low activity similar to that of the Maunder Minimum from 1645 to 1710. The reduction in solar activity now being observed will result in temperatures returning to the levels of the mid-19th century at best, with the possibility of revisiting the lows of the 17th and 18th centuries. Peak summer temperatures may not change much but the length of the growing season will shorten at both ends, playing havoc with crop yields.

The notion of global warming has resulted in an enormous mis-allocation of resources in some Western societies but we can be thankful to it for one thing. If it had not been for the outrageous prostitution of science in the global warming cause, then the field of climate would not have attracted the attention that has determined what is actually happening to the Earth's climate. Humanity would otherwise be sleepwalking into the severe cold period in train.

Global warming due to carbon dioxide is of no consequence and the world is cooling anyway.

Let's have what China is having in CTL

There has been a recent burst of activity in China, with 20 coal conversion projects approved by 2013. CTL production will rise to near 1.0 million barrels per day by 2020, requiring 180 million tonnes of coal per annum. This is just about what Australia needs. China also has 20 coal-to-methane projects in train which will require a further 200 million tonnes of coal per annum. In total that amount is about what Australian coal mines produce each year. All this Chinese activity is that Australia need not be tentative in relying upon CTL to solve our liquid fuel security problem. The technology has been proven and re-proven.

The other thing that China is doing to improve its liquid fuel security is stockpiling oil. At one point in 2014, it was estimated that China was placing 1.4 million barrels per day into storage. China now has stocks of at least 700 million barrels, equivalent to that of the US Strategic Petroleum Reserve, which will be useful in any war it might start.

Other issues

CTL plants produce a pure CO_2 stream from the synthesis gas cleanup stage. This may prove to be a valuable resource in that algal growth for many species is optimised at around 10 per cent CO_2 compared to the atmospheric content of 0.04%. Each barrel of fuel produced has 200 kg of CO_2 as a byproduct. This will make 100 kg of algae which in turn will yield 0.34 barrels of biodiesel. Thus a 10,000 barrel per day plant could make a further 3,400 barrels per day of biodiesel. This is a big potential increase in the project fuel yield.

Water use is insignificant in CTL production. Growing sorghum under irrigation requires 1,000 tonnes of water to yield one tonne of grain worth $200. By comparison, through a CTL plant it takes four tonnes of water to produce one tonne of diesel worth $940. The revenue yield of water is 1,200 times greater through a CTL plant than it is through grain.

At two barrels to the tonne, one tonne of coal produces 318 litres of fuel. The Mazda 3 has a fuel economy of 17.5 kilometres to the litre. Thus one tonne of coal will fuel a Mazda 3 for 5,580 km.

For an average grain farm in Australia to supply its own biodiesel from canola grown on the farm, 20 per cent of the area of the farm would have to be devoted to growing canola. Thus for the farm sector in Australia to supply its own fuel, farm production would fall 20%. Supplying any other part of the economy with fuel would mean a further significant fall in food production. Thus biofuels are a distraction on the path to transport fuel security.

Ultimately when fossil fuels are exhausted, the only viable source of energy to maintain civilisation at a high level is nuclear. The optimum nuclear power process is thorium molten salt reactors which are expected to have a cost of 3 cents/kWh. At that price, hydrogen could be produced by electrolysis at the equivalent energy cost of US$60 per barrel. From that, hydrocarbon fuels might be produced using organically grown carbon for US120 per barrel and civilisation could continue at a high level indefinitely.

The Gillard Government imposed the Resource Rent Tax (RRT) on onshore oil and gas production in 2011. The current Federal Government retained the tax because it would like to get an extra income from the recently built onshore LNG projects based on coal seam gas. However those projects will have single digit internal rates of return, or even zero rates of return, and no RRT will be paid. The tax will suppress onshore conventional oil exploration as it takes the total tax take up to about 70%. Onshore oil exploration, and ultimately onshore oil production, is collateral damage from a tax which isn't going to raise any funds from its primary target. It would be a pity if the nation were lost due to a tax that didn't raise any funds.

5

CHINA'S COMING WAR

Introduction

After the collapse of most of the world's communist states in 1990, the world appeared to have entered a period of permanent peace. The Stanford University-based political scientist Francis Fukuyama called it "the end of history" in which democracy and free market capitalism would become the final form of human government. In response to Fukuyama's 1992 book[1], the Harvard historian Samuel Huntington penned an article entitled "The Clash of Civilizations?" which was expanded into a book in 1996 entitled "The Clash of Civilizations and the Remaking of World Order".[2] Huntington argued that now that the age of ideological conflict had ended, the world's normal state of affairs of civilisational conflict would reassert itself. His book concentrated on the "bloody borders" between Islamic and non-Islamic communities. His insights seemed particularly prescient after the Islamic attacks within the United States on 11th September 2001.

Apart from the Islamists, there is another civilisation that is unhappy with the world as it is and wants to change it. The 2001 attack overshadowed one the previous year, far away in the South China Sea. On April 1, 2000, a Chinese jet fighter backed into a US reconnaissance aircraft flying at 22,000 feet and 70 miles southeast of Hainan Island. The Chinese jet crashed, the US aircraft landed on Hainan Island where the crew of 24 were held captive until April 11. Tension between China and the United States mounted as the days of captivity passed and it seemed at the time that the next step for the United States would have been to impose trade sanctions on China, which backed off at the last moment.

The civilisational clash with Islam has settled down for the present to desert skirmishes and drone attacks on the unhappy. The civilisational

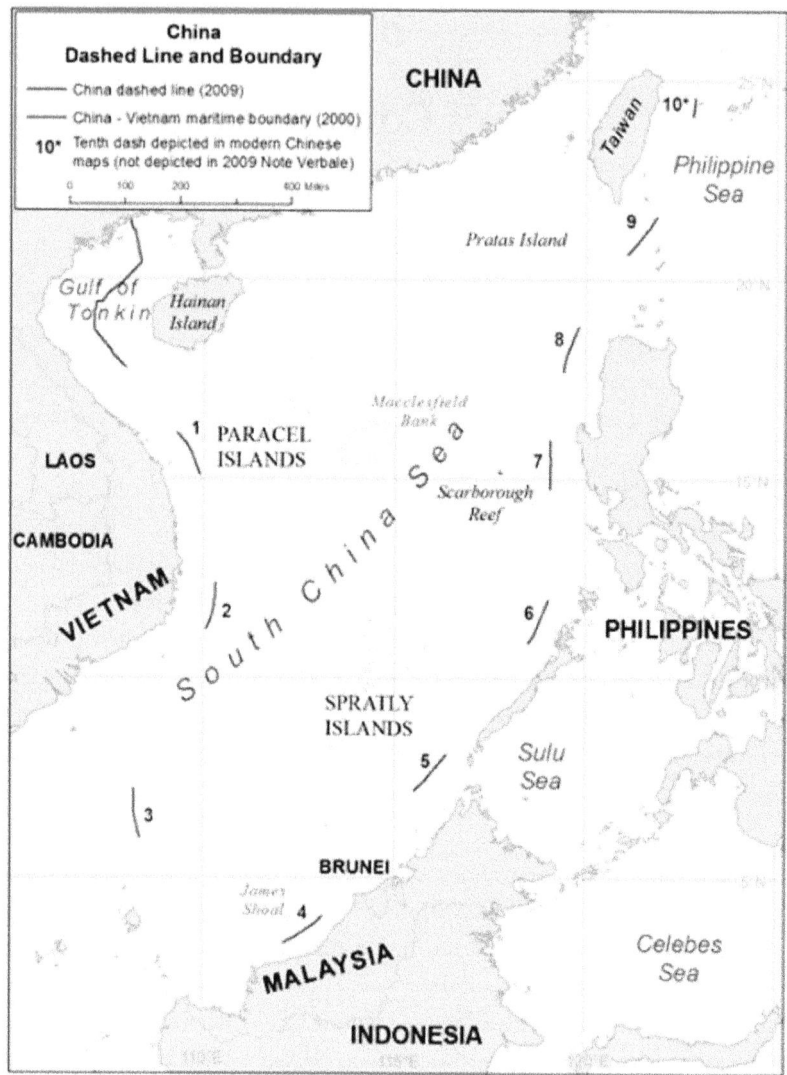

Figure 25: China's Claim Area in the South China Sea

This map shows China's "nine-dashed claim" area of the South China Sea, originally drawn up by Nationalist officers in 1947. A tenth dash was added in 2009 just west of the Japanese island of Yonaguni in the Yaeyama Islands.

clash with the unhappy Chinese will be something different altogether. As Pentagon strategist Edward Luttwak pointed out in his 2012 book[3] "The Rise of China and the Logic of Strategy", there are many parallels between China and Germany in the lead up to World War I. Germany at the time thought that it was not given enough respect. All other major powers had empires. Germany had been late to the party and picked up the scraps around the planet, such as the north-eastern third of the island of New Guinea and a number of Pacific Ocean islands. Germany, at the time, felt compelled to go to war. It planned on a quick war but it did not turn out that way.

One hundred years later China is bent on following the example of Wilhemine Germany. It was late to the industrialisation party but made up for this with a ferocious rate of capital investment. The Chinese have traditionally seen themselves as the most civilised people on the planet. They also prefer that other nations be deferential to them in a hierarchical arrangement. Their intrinsic view of the world was confirmed by the Global Financial Crisis of 2008, during which the Europeans begged to be bailed out of their predicament with Chinese money. That would have sealed the deal in terms of their contempt for foreign cultures that are far more self-indulgent than China's. In fact China's harsher tone dates from 2008.

Some have seen this war coming well in advance. In 2005, Robert Kaplan wrote an article entitled *How We Would Fight China* in which he notes that China will approach the war "asymmetrically, as terrorists do. In Iraq the insurgents have shown us the low end of asymmetry, with car bombs. But the Chinese are poised to show us the high end of the art."

In terms of gaining an empire their efforts are far more pitiful than what Germany had gained prior to World War I. China is in the process of attempting to seize the South China Sea as far south as the Natuna Islands, part of Indonesia. Their claim bumps up against the coast of Borneo. The area has been almost completely uninhabited because there was nothing worth staying for. There were no fishing settlements in the area so the fishing cannot have been that attractive. In terms of oil and gas potential there may be some out from the coast of Vietnam on the continental shelf. The rest of the area is deep water with coral reefs and carbonate platforms in the same style as the Bahamas Platform east of Florida. In short, there are no natural resources worth losing blood over. The claim is purely political.

The trouble for China is that now that they have upped the ante by stating that they will enforce their claim, it is therefore difficult for them to back down without losing the respect they crave in the first place. So this will end in tears, but for whom?

The Chinese force structure is based on area denial, with a swarm of missile-firing high speed catamarans at one end of the force spectrum and DF-21D ballistic anti-shipping missiles at the other. The DF-21D missiles, with a range of 2,700 kilometres, are designed to sink United States aircraft carriers. China has also stepped up its hacking of utilities and other public infrastructure in the United States, laying the groundwork for a potential "cyber-Pearl Harbour". Ideally, for the Chinese, they would like to sink an aircraft carrier and then call a halt to hostilities. The United States' influence would shrink back to Hawaii and then China would be able to do whatever it wanted in Asia.

China does not yet have all the weapons it desires for it future conflicts. They are still having trouble making nuclear powered submarines and jet fighter engines. They may not feel the need to wait for their technological abilities in those areas to catch up.

There was an idea in the mid-19th century that trade promoted peace. It was believed by some that once countries realised they did better trading with one another than fighting, peace and goodwill amongst nations would prevail. This was the theme of two peace conferences in the UK in 1853. That was also the year that Commodore Matthew Perry visited Japan and forced it to open up to trade and modernisation. A scant 20 years later, the Japanese Government discussed attacking Korea. Japan attacked China first, in 1895, took over Korea in 1905, then annexed it in 1910. Japan just kept on attacking as the decades passed and left bitter memories throughout the region. And some of the victims love keeping those memories alive. Today 30 per cent of Chinese prime time television is devoted to movies about the Japanese invasion of the 1930s.

One problem with trade and rapid economic development is that it tends to make people over-confident. Bethmann Hollweg, chancellor of Germany at the outbreak of World War I, confessed subsequently that Germany had over-valued her strength. 'Our people', he said, 'had developed so amazingly in the last twenty years that wide circles succumbed to the temptation of over-estimating our enormous forces relative to the rest of the world.' That sounds exactly like China 100 years later. And there is the problem of when economic growth falters. In 1982, Argentina attacked the Falklands to provide some patriotic legitimacy to the generals' regime when its economy faltered.

China continues to prepare for war. It has done a good job of convincing its neighbours that one is coming. Over 60 per cent of people in countries bordering the South China Sea fear Chinese aggression and expect war. The Chinese continue to convince themselves that war is inevitable. A Chinese Government film made in late 2013 made for consumption within the party and the military, *Silent Contest*, began with these words:

> The process of China's achieving a national renaissance will definitely involve engagement and a fight against the U.S.' hegemonic system. This is the contest of the century, regardless of people's wishes.

The basis of the film is that the US used cultural engagement with the Soviet Union to destroy that country and is also using cultural engagement to contain and divide China. The fact that China considers itself to be involved in a titanic "contest of the century" with the U.S. would be news to most Americans. But the Chinese are not content with having lifted themselves out of poverty by making geegaws. They crave the respect that only a resounding military victory can bring.

China's motivations

China will start its war for a number of reasons:

1. Regime legitimacy

Very few people in China believe in communism anymore, including almost all of the 80 million members of the Chinese Communist Party. The party itself is now a club for mutual enrichment. The legitimacy of the party ruling China is derived from the notions that democracy does not suit China and that the party is the organisation best placed to run the country. The latter is based on an ongoing improvement in conditions for the bulk of the population. In the absence of economic improvement, some other reason must be found for the population to rally around the party's leadership. This may explain the sudden base-building that started in the Spratly Islands in October 2014.

China's public debt grew from US$7 trillion in 2007 to US$28 trillion in 2014. This is on an economy of US$10 trillion per annum. A high proportion of the economic growth of the last seven years is simply construction funded by debt. The real economy is much smaller.

The Chinese government is likely to see the contracting economy and realise that issuing more debt won't have an effect on sustaining economic activity. Thus the base-building in the South China Sea was accelerated in 2014 to allow the option of starting their war. This is a life and death matter for the elite running the party. They are betting the farm on this. If this gamble does not work out then there is likely to be a messy regime change.

2. Chosen trauma

Japan treated the Chinese as sub-humans during World War 2. Before that, Japan starting mistreating China by attacking it in 1895, not long after they started industrialising themselves. That was followed by Japan's 21 demands on the Chinese state in 1915. The Nationalist government in China started observing National Humiliation Day in the 1920s on the anniversary of the date of Japan's 21 demands. The Mukden Incident followed in 1931 and China's and Japan's start to World War 2 in 1937.

During the poverty of the Mao years, the Japanese were forgiven for World War 2. Mao and Deng were pragmatists and said that Japan

China's Coming War 121

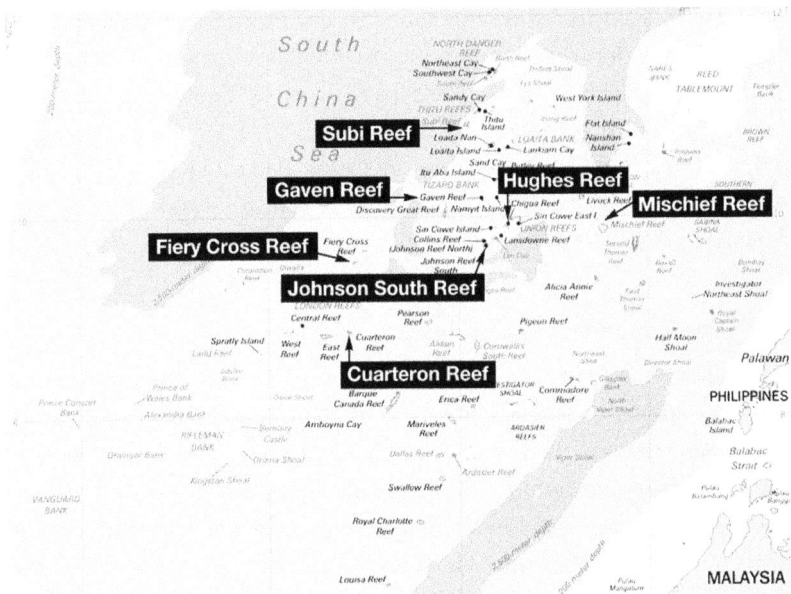

Figure 26: Chinese Base-building in the Spratly Islands (image courtesy of Victor Robert Lee)

Starting in late 2014, China mobilized a large number of cutter-suction dredges to reefs and atolls in the Spratly Islands, turning seven of them into artificial islands. All of these islands include a harbour. An airfield has been built on Fiery Cross Reef and airfield construction has begun on Subi Reef. At least four of the islands now have forts build on them. The forts, with reinforced concrete walls three metres thick, include hexagonal flak towers set apart from the corners. The locations of the bases are:

Fiery Cross Reef	9° 32' N, 112º 53' E
Gaven Reef	10° 12' N, 114º 13' E
Subi Reef	10° 55' N, 114º 05' E
Johnson South Reef	9° 43' N, 114º 17' E
Cuarteron Reef	8° 52' N, 112º 49' E
Hughes Reef	9° 54' N, 113º 30' E
Mischief Reef	9° 54' N, 115º 31' E

couldn't be punished forever. China's recent prosperity has allowed the indulgence of Japan-hating to be resurrected as a form of state religion. National Humiliation Day is observed again as 18th September. The party has directed that television take up the theme of Japanese aggression. There are at least 150 museums in China dedicated to the Japanese aggression of World War 2. The regime generates and sustains anti-Japanese sentiment to give it the option to go to war.

3. Being recognised as number one

The Chinese are a proud nation. They actually resent the fact that the United States is considered to be the number one nation on the planet. China also realises that to be recognised as number one, they have to defeat the current number one in battle. This is why it won't be just creeping increments in Chinese aggression. They need a battle for their own psychological reasons. This means that they will attack the United States at the same time that they attack Japan. Because surprise attacks are more successful, it will be a surprise attack on US bases in Asia and the Pacific and perhaps well beyond. This most likely will include cyber-attacks on US utilities and communications.

China has structured its armed forces for a short, sharp war. Of any country on the planet, they are possibly the most prepared for war. They have one year of grain consumption in stock and even a strategic pork reserve. They have just filled up their strategic petroleum reserve of about 700 million barrels. China's attempt at seizing the South China Sea has nothing to do with securing resources or making their trade routes secure. Some western analysts have projected those notions onto China to rationalise what China is doing. The Chinese themselves have not offered these excuses. To China it is all about territorial integrity, which is sacred and not the profane stuff of commerce.

4. Humiliating the neighbours

The importance of the Spratly Islands and the Chinese nine-dash claim is that it divides Asia. China claims that the whole of the sea within its claim is Chinese territory, not just the islands. When China gets around to enforcing that claim, foreign merchant vessels and aircraft will have to apply for permission to cross it. Non-Chinese warships and military aircraft will not be allowed to enter it. The Chinese claim extends to 4° south, almost to the equator.

The worst affected country will be Vietnam which will be bottled up to within 80 kilometres of its coast. Japan realises that its ships from Europe and the Middle East will have to head further east before heading up north through Indonesia and east of the Philippines. Singapore will be badly affected because the passing trade will drop off. Japan will become quite isolated because its aircraft will have to head down through the Philippines to almost the equator before heading west. China ranks the countries of the world in terms of their comprehensive national power which the Chinese consider to be the power to compel. This is a combination of military power, economic power and social cohesion. When it is enforced, the nine-dash claim will do a lot of compelling of China's neighbours.

5. Strategic window

Chinese military writers see a window of strategic opportunity for China early in the 21st century though they haven't publicly outlined the basis for that view. But we can make a good stab at it. Firstly, an air of inevitability is important in winning battles. While China is perceived to have a strong, growing economy that is crushing all before it, that perception of inevitability rubs off on China's military adventures. To use that perception, China has to attack before its economy contracts due to the bursting of its real estate bubble. This explains the current rush to build the bases in the Spratly Islands.

Another problem for China is that its aggression and increased military spending has caused its neighbours to rearm and form alliances. China is better off attacking before its neighbours arm themselves further.

Another consideration is the US presidential electoral cycle. President Obama has proven to be a weak president and the Chinese might rather attack before he is replaced. President Obama has made the right noises though about Chinese irredentism and the coming war remains quite popular in the US military in that the different services are jockeying for position, which means they have official blessing to the highest level. But he also does not want to be remembered as having had a major war start while he was president. President Obama does have some inconsistent policies which aid China though in that while a strong economy is needed to fight China, his administration is doing its best to choke the US economy with carbon dioxide-related regulations. The two ends are mutually exclusive.

President Obama spent a period of his childhood in Indonesia and would have heard a lot of anti-Chinese sentiment (the Chinese were and are more successful merchants and shopkeepers) in those formative years. As with Valerie Jarrett's childhood in Iran, this will affect policy.

6. Great-State autism

This is a term created by the strategist Edward Luttwak to describe the fact that China is seemingly oblivious to the effects of its actions on its neighbours. China sees itself as the centre of the world and purely through the lens of its own self-interest. This has the practical result that China could not perceive of things not going the way it wants them to. Luttwak also considers that the Chinese overestimate their own strategic thinking. He says that China doesn't have a strategy so much as a bag of stratagems, most of which involve deception.

7. President Xi Jinping

While preparation for this war started in the 1980s, the recent ramp up in aggression has been at the direction of President Xi, who, in his formative years as a party apparatchik, was impressed by how the war with Vietnam in 1979 was used to consolidate power in the politburo. President Xi has accumulated more power than any Chinese leader since Deng Xiaoping. He is using an anti-corruption campaign to purge political opponents.

Chinese leaders are supposed to only rule for ten years before standing aside. Just two years into his presidency, Xi's supporters have raised the possibility of resurrecting the position of chairman of the party (abolished by Deng to stop another Mao) so that Xi could continue to rule from that position. President Xi is a nasty piece of work who has been toughened up by his life experiences. At the age of 15, he was sent to live and work with peasants in the yellow earth country after his father was purged. His accommodation was a cave. His step-sister suicided due to his father's oppression by the Red Guards.

Japan and the United States

Japan sees this war being thrust upon it and is approaching it with a great deal of foreboding. It sees it as being inevitable, though Prime Minister Abe did ask to meet President Xi in Indonesia in early 2015. President Xi intends to kill many tens of thousands of Prime Minister Abe's countrymen, so the meeting was strained. Prime Minister Abe recently addressed a setting of the US Congress, part of his doing the rounds to make sure everyone is on the same page with respect to absorbing and repelling the Chinese attack.

The United States believes that a rules-based world order needs to be maintained for global security and prosperity, including its own prosperity because that relies upon world trade to a large extent. So for the United States, this war will be about preserving access to the global commons. The US military establishment has not kept the public up to date with all of China's preparations for war probably because they do not want to be perceived to be causing escalation. But the US military is in no doubt that China will start a war. The main unknown is the timing.

Chinese aggression has been a godsend to the US Navy which had lacked a credible threat and had faced ongoing shrinkage. There is a tendency to overstate the efficacy of enemy weapons systems. The Chinese would have read the US Navy reports on their weapons systems which would have emboldened them further.

Chinese threat signalling

In 2013, the Center for the Study of Chinese Military Affairs at the Institute for National Strategic Studies in Washington D.C. published a paper by Paul Godwin and Alice Miller entitled "China's Forbearance Has Limits: Chinese Threat and Retaliation Signaling and Its Implications for a Sino-American Military Confrontation". The paper is quite timely because it should be possible to predict the sequence and timing of China's move towards war. By the authors' research of China's conflicts with its neighbouring countries after World War 2, Chinese threat signalling should follow four stages in a conflict:

First, systematic integration of political and diplomatic action with military preparations as the signaling escalates through higher levels of authority. These preparations are normally overt and designed to "deter the adversary from the course of action Beijing finds threatening."

Second, China states why it is justified in using military force should this prove necessary. The message targets both domestic and international audiences. "In essence, Beijing declares that China confronts a serious threat to its security and interests that if not terminated will require the use of military force." Third, China begins asserting that the use of force is not Beijing's preferred resolution to the threat, but one that will be forced upon it should the adversary not heed the deterrence warnings sent. The signaling strategy seeks to grant China the moral high ground in the emerging confrontation. "Such argument supports China's self-identification as a uniquely peaceful country that employs military force only in defense when provoked by adversaries threatening China's security or sovereignty." The authors suggest China believes that asserting the moral high ground in a fight can ease the international response to any military action it might take and thus reduce the political costs of employing military force.

Fourth, Beijing emphasises that China's forbearance and restraint should not be viewed as weakness and that China is prepared to employ military force should that be necessary. These four signals, or check lists for war, reflect a basic pattern China has demonstrated since its first signaling in 1950 when China sought to deter US forces from crossing the 38th parallel into North Korean territory.

How the war will be conducted

There will be two main theatres of operation: the East China Sea north of Taiwan and the South China Sea west of the Philippines.

China claims sovereignty over the Senkaku Islands (last occupied by the Japanese about 100 years ago) and the entire Ryukyu chain from the Yaeyama Islands at the southern end to, and including, Okinawa in the north. If it is going to seize the Senkaku Islands, it might as well seize the Yaeyama Islands at the same time. To that end, China is building up a military base in the Nanji Islands about 300 kilometres west of the Senkakus. This includes a 10-pad helicopter refuelling base which suggests that the initial assault will be led by helicopters overflying Japan's coast guard vessels around the Senkakus.

China has a substantial fishing vessel fleet and merchant shipping totalling 70 million tonnes. It has been using its fishing fleet to harass the Japanese coast guard around the Senkakus and as far east at the Osagawa Islands, which includes Iwo Jima. This suggests that fishing vessels could be used to land Chinese special forces to widely attack Japanese bases that would normally be considered to be well back from the front line. These forces would be used sacrificially to cause maximum mayhem to dispirit the Japanese defence. In the north, the Chinese approach would be to seize and hold against the Japanese and US counter attack. In the south, the chief of a fishing corporation on Hainan Island has outlined an approach that China is likely to rely on:

> If we put 5,000 Chinese fishing ships in the South China Sea there will be 100,000 fishermen. And if we make them all militiamen, give them weapons, we will have a military force stronger than all the combined forces of all the countries of the South China Sea. Every year, between May and August, when fishing activities are in recess, we should train these fishermen/militiamen to gain skills in fishing, production and military operations, making them a reserve force on the sea, and them to solve our South Sea problem.

In Hainan Province on the northern edge of the South China Sea there are more than 23,000 fishing vessels available for this purpose. To sink all the fishing boats that China will deploy will require a stock of thousands of air-to-surface missiles.

For larger ships, China has instituted military standards that new civilian ships have to be built to, covering five categories of vessels: container, roll-on/roll-off, multipurpose, bulk carrier and break bulk. China had about 172,000 civilian ships and small craft at the end of 2014. More than 11,000 were dedicated to inshore transport and around 2,600 performed ocean transport.

As a quid pro quo for China supporting it over Crimea and the eastern Ukraine, Russia is in the process of selling China six battalions of S-400 surface-to-air missile systems with a range of 400 kilometres. Each battalion consists of a command post, radar and eight launcher trucks with four missiles each. The missiles have a maximum velocity of 4.8 kilometres per second. The six battalions amount to 192 missiles, not including reloads. Operating practice is to fire two missiles at one target to increase the hit probability. So these systems put 96 US and allied aircraft at extreme risk at up to 400 kilometres from the Chinese coast or their island bases.

A year ago, the commanding officer of the US Marines in Japan said that retaking the Senkaku Islands from China would be relatively easy and that he didn't need the US Army's help in the matter. He was even concerned that the US Army's helicopters might suffer corrosion while embarked at sea. The Senkakus are 300 kilometres from the Chinese mainland and retaking them now will be somewhat more difficult thanks to the S-400 systems.

In the South China Sea, China is building seven massive forts and three airfields. The forts are designed with flak towers standing out from the corners so that each tower has at least a 270° field of fire. The forts seem to be designed to take a large amount of punishment and hold out until they can be relieved. China wins if it is still in the possession of these forts by the end of the war.

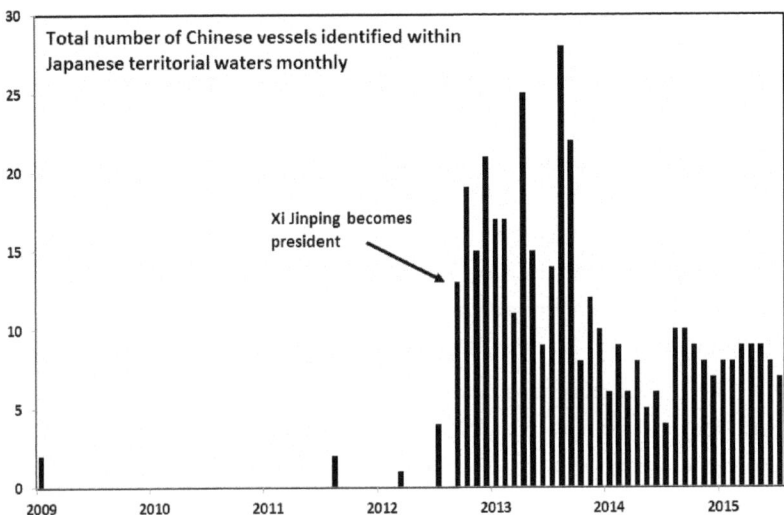

Figure 27: Trends in Chinese Government Vessels in the Waters Surrounding the Senkaku Islands

On 8th December 2008, two Chinese government vessels made a sudden intrusion into Japan's territorial sea surrounding the Senkaku Islands. Despite repeated calls by patrol vessels of the Japan Coast Guard to leave the area and strong protest lodged against China through diplomatic channels, the vessels hovered and drifted inside Japan's territorial sea for some nine hours until the evening of that day. The incident made clear China's new position concerning the Senkaku Islands, one that had never been observed before: Chinese government vessels intrude into Japan's territorial sea with the clear intention of violating the sovereignty of Japan, attempting to change the status quo through force or coercion

After a collision of a Chinese fishing boat into Japan Coast Guard patrol vessels in Japan's territorial sea surrounding the Senkaku Islands on September 7, 2010, Chinese government vessels started to sail the waters surrounding the Islands more frequently than before. In August 2011, two Chinese government vessels intruded into Japan's territorial sea surrounding the Senkaku Islands, preceding one in March 2012 and four in July that year. The rate of intrusions increased dramatically with the accession of Xi Jinping to China's presidency. For the twelve months up to July 2015, the rate of state-directed intrusions is continuing at about nine per month.

China is likely to start the war in the south with attacks on other countries' bases in the Spratly Islands and US bases in the region, as far east as Guam. A long war will be bad for China in that the run down to the Spratly Islands from Hainan Island is very exposed, both for ships and aircraft. Vietnam has been upgrading its radars and hopefully all the non-Chinese combatants will be sharing targeting information. US AWACS over the Philippines will be able to track Chinese targets handed over from Vietnam. Singapore is likely to operate its F-15s out of Cam Ranh Bay. Chinese aircraft that survive the run down will be at the end of their range by the time they get to the Spratly Islands.

The US Marines have taken up a number of bases in the Philippines with the intention of mounting the attack that will remove the Chinese from their newly constructed forts. A number of US weapons systems, such as the USS Zumwalt, may have to be rushed into service to that end.

In the bigger picture, Japan and China will try to blockade each other, mostly with their submarine forces. Japan's navy has a qualitative edge over China and is most likely to win the blockade battle.

Industry throughout Asia will be badly affected by the war but Chinese industry in particular is likely to grind to a halt quickly and this will eventually cause social disruption. The longer the war goes on, the worse China's relative position becomes. Meat will disappear from the Chinese diet. Unsold soybeans will pile up in US warehouses.

The removal of the Chinese bases in the Spratly Islands will allow a peace settlement with whoever ends up running China. It will be one of the most pointless, stupid and destructive wars in history, but that is what is coming.

We are very likely to win this war. Starting in the 1920s, it was realised that Japan would one day attack US interests in the Pacific. War Plan Orange to defeat Japan was formally adopted by the Joint Army and Navy Board in 1924. It was a good plan and was successfully executed in World War 2. Nearly one hundred years later, Air Sea Battle was adopted as the plan for defeating China. In early 2015 that name was changed to

Joint Concept for Access and Manoeuvre in the Global Commons, to make it more inclusive so that the US Army could have a role. It is also likely to be a good plan. We can tell that some of the holes are being filled in by things such as the basing choices in the Philippines. Technological development has aided the defence, and that is what we will be doing. The forts that China is building in the Spratly Islands will soak up a lot of ordnance but for their ships and planes, the South China Sea is a natural kill box. There will be a lot of surprises in this war but we will prevail.

Australia's role

Australia's Quisling Central is something called the Australia-China Relations Institute (ACRI) based at the University of Technology Sydney. ACRI is led by Bob Carr, the former premier of New South Wales. It was launched on May 16, 2014 and the Australian Foreign Minister, Ms Julie Bishop, was there to fawn over the Chinese. It is amusing to note that she urged "China to work constructively and cooperatively to resolve tensions in our region" when China itself is the source of 100 per cent of those same tensions. ACRI was possibly set up with one purpose – winkling Australia out of our defence alliances with Japan and the United States ahead of the Chinese attack on those countries. Everything ACRI produces is agitprop to that end.

Thus, seven months after founding, ACRI came out with a survey it commissioned saying that 71 per cent of Australians thought that Australia should stay out of any war between China and Japan. We don't know whether that number is true or not depending upon the wording of the questions asked. No matter what the words used, it is push-polling. ACRI is trying to alter the political landscape in the aggressor's favour. The poll is another confirmation that China is planning to attack Japan – why otherwise would they do it?

The benefactor that funded the founding of ACRI is a Mr Huang Xiangmo of the Shenzhen-based real estate developer Yuhu Group. He and his group funnelled $350,000 to the ALP in 2012-13. Perhaps it was coincidence, but perhaps not, because in 2013 the Gillard Government started a thing called the Australia-China Security Dialogue. While the Chinese political system is endemically corrupt, we try to be transparent. Thus one can find reference to the donation made by Mr Huang to our Minister for Social Security, Scott Morrison, of a Mont Blanc pen valued at $330.

Bob Carr is now Australia's wrinkliest Quisling following the death of Malcolm Fraser. The last fruit of Fraser's poison pen was a tome called *Dangerous Allies* in which he called for Australia to cut its close ties to the United States. The benefit of Malcolm Fraser's opinion is that he was an extremely reliable counter-indicator. Just as global warming was a litmus test for our politicians – the ones who believed in it were the easily deluded fools, Fraser had consistently been on the wrong side of history. Now this reliable indicator has been taken from us, and if there is any justice in the next world, his soul will be in eternal damnation.

Senator David Johnston's dismissal as Australia's Defence Minister came soon after he said that he wouldn't trust ASC (formerly Australian Submarine Corporation) to build a canoe. The ASC sheltered workshop wanted to be paid $3 billion a copy to build submarines. The Japanese build them year in and year out for $800 million a copy. Surely an attempt to be an effective Defence Minister shouldn't be a sacking offence in an ideal world.

The Senator had said something far more disturbing earlier in the year, something that would qualify as a sacking offence. In June 2014, when, in response to the question, *"Does the ANZUS alliance commit Australia or not if the United States is in a conflict in our region?"* he replied, *"I don't believe it does."*

That answer got Quisling Central so excited that they published a booklet based on it. The booklet, entitled *Conflict In The East China Sea: Would ANZUS Apply?* is written by academics Nick Bisley and Brendan Taylor. The ACRI booklet waffles on for 80 pages but it can be summarised in its first recommendation:

> The principal challenge for Australia lies in maintaining maximum freedom of policy manoeuvre in the event conflict erupts in

the East China Sea. This means ensuring that Australia does not overcommit too soon, thus taking a position in which it unnecessarily pays a price with Beijing. For Canberra the main piece of policy preparation lies in managing the expectations of the US and Japan in the event of conflict.

Note that Bisley and Taylor do not answer the question they ask in their title. So let's answer that question ourselves by reading the original document (included as Appendix 1). Article IV of the ANZUS Treaty includes the sentence:

> Each Party recognizes that an armed attack in the Pacific Area on any of the Parties would be dangerous to its own peace and safety and declares that it would act to meet the common danger in accordance with its constitutional processes.

That means that if the United States was attacked in the Pacific region then Australia and New Zealand would be obliged to render military assistance to the United States. To make sure that there is no doubt about the meaning of Article IV, Article V was included which states:

> For the purpose of Article IV, an armed attack on any of the Parties is deemed to include an armed attack on the metropolitan territory of any of the Parties, or on the island territories under its jurisdiction in the Pacific or on its armed forces, public vessels or aircraft in the Pacific.

There is no ambiguity there. If China sinks an American ship or shoots down an American aircraft, military or otherwise, and the United States declared war on China, Australia would be at war with China too.

Senator Johnston's answer in mid-2014 abrogated the one treaty that Australia's whole defence posture is based on. It may have been a moment of befuddlement or carelessness, but such carelessness can get a lot of people killed. A good example from recent history is April Glaspie. She was the United States ambassador to Iraq and told Saddam Hussein "*We have no opinion on the Arab-Arab conflicts, like your border disagreement with Kuwait.*" What is not specifically forbidden in such matters is allowed, so

Saddam Hussein took Ambassador Glaspie's words as a green light to invade Kuwait. We are still dealing with the aftermath 25 years later.

And so it is with the East and South China Seas. Any minister of the crown attempting to be nuanced and sophisticated with respect to Australia's obligations under the ANZUS treaty is encouraging China to be more assertive and start the war they want. And as Confederate general Nathan Forrest said "War means fighting, and fighting means killing."

A couple of ministers of the crown, Prime Minister Tony Abbott's presumptive heirs, have been careless in their comments if Australia's aim is to help preserve the peace in Asia. Amongst other things, Communications Minister Malcolm Turnbull has written:

> We should seek to ensure that the Americans, unlike the Spartans, do not allow their anxiety about a rising power to lead them into a reflexive antagonism that could end in conflict.

Talk about getting the situation the wrong way round.

Foreign Minister Ms Julie Bishop, a self-confessed very good friend of Malcolm Turnbull of 20 years standing, has said:

> The United States has long been the single greatest power in the Pacific, in Asia, in fact globally. ... But we recognize that there are other countries that are emerging as stronger economies, other countries are building up their militaries. ... So we are in a very different world. It's a changing landscape and our foreign policy must be flexible enough and nimble enough to recognize that changing landscape.

Ms Bishop's words, uttered as Foreign Minister, and so nuanced and sophisticated, are exactly the sort that will get so many people killed. Prime Minister Tony Abbott has spoken of Japan being "Australia's best friend in Asia". That is the kind of statement that helps keep the peace in our time.

China is on a path to attacking a number of its neighbours and subjecting them to humiliation in the subsequent peace. Would anybody want a friend like that? The Quislings amongst us would have us abandon our relationship with the United States and hitch our wagon to the Red Star of Asia, which is supposedly rising. Let's examine that claim first.

The Chinese economy took off after their accession to the World Trade Agreement in 2001. China became the world's preferred subcontractor. Electronic components made in Japan, Korea and Taiwan are imported into China and put in plastic cases. The salad days of China's export-driven growth ended in 2006, when exports as a percentage of GDP peaked at 39.1 per cent. That has subsequently fallen to 26.4 per cent. But the world can only take so much of China's exports and that point has been reached. China's market share of global trade reached 12 per cent in 2011 and has stalled at about that level. That follows railway freight in China, which doubled between 2003 and 2011 but has stalled since. Previous premier Wen Jiabao said in 2013 that China's growth is "unbalanced, uncoordinated and unsustainable". Another thing about China's growth: it may be illusory.

We don't have to spend too much time on economic statistics to divine China's future. All we have to do is note what the Chinese themselves are doing, which is leaving. As John Lee has noted, the richest 1 per cent of households (2.1 million out of a total of about 520 million households) own 40-50 per cent of the country's total real estate and financial assets.[4] This is the result you would expect from a state-sanctioned kleptocracy. These wealthy people are voting with their feet. In a survey in 2014 of almost 1,000 Chinese, each worth over $16 million, nearly two-thirds had made arrangements to leave the country permanently or were planning to do so. This group is particularly well-informed on China's prospects, with 90 per cent of the 1,000 polled being officials or members of the Chinese Communist Party. These are people who have stolen what they can and now think that wealth preservation is more important than hanging around to steal some more. When the rats are leaving the ship, why would our Quislings have us clamber aboard a junk that is settling in the water?

But it is worse than that. The sudden increase in Chinese aggression in the Senkakus in late 2012 was about the time that Xi Jinping became

General Secretary of the Chinese Communist Party. He is now also President of the People's Republic of China, Chairman of the Central Military Commission and Chairman of the National Security Council. Tony Abbott was quite pleased that President Xi, in his address to the Australian Parliament after the Brisbane G20 meeting, said that China was on track to become a liberal democracy by 2050. At home in China, he is doing the opposite. In a number of edicts, he has tacked hard left politically and is railing against foreign influences in Chinese society. Under the guise of fighting corruption, he has instituted a reign of terror equivalent to the Stalinist purges of the 1930s. Or perhaps they are just recycling more recent Chinese history. To quote long-time China watcher Anne Stevenson-Yang in late 2014:

> What's really going on is an old-style party purge reminiscent of the 1950s and 1960s with quota-driven arrests, summary trials, mysterious disappearances, and suicides, which has already entrapped, by our calculations, 100,000 party operatives and others. The intent is not moral purification by the Xi administration but instead the elimination of political enemies and other claimants to the economy's spoils.

Given that President Xi is quite happy to kill thousands of his countrymen to consolidate his political position, the lives of foreigners would be the merest trifle. We should not be trying to get closer to these people. Selling them some iron ore is enough. But our Quislings would have us abandon the United States for the world's largest mafia operation? And some who should know better, including Australia's Foreign Minister, have said some unfortunate things.

China's economy is not bigger than that of the United States. The relative size of their economies is possibly most accurately captured by their respective oil consumption figures, China's 10 million barrels per day versus the United States' 18.5 million barrels per day. China's economic effectiveness is diluted by hundreds of millions of rural peasants who make only a small contribution to the economy. But no matter what economic facts and trends portend, throwing our lot in with China will end badly for us, and could very well be the end of us.

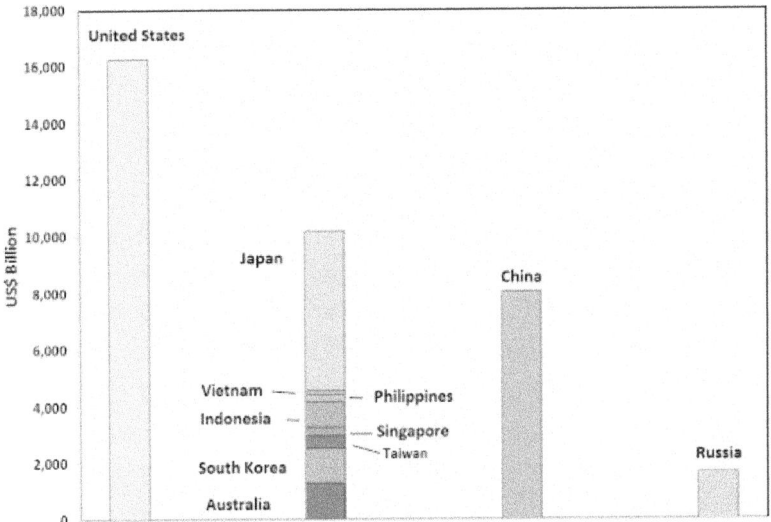

Figure 28: Relative Size of Economies in East Asia
Given that an attacker should have three times the force available to a defender, Japan alone could hold off China. Combined with the other countries and the United States, the economies of the potential allies against Chinese irredentism are three times larger than China's economy.

That said, there is a historical symmetry to the debate. Prior to World War II, Prime Minister Robert Menzies was nicknamed "Pig Iron Bob" because he exported scrap iron to Japan that was used to make bombs to kill Chinese. Now the Left in this country is saying that our iron ore exports to China are so important that we should abandon our relationship with the United States, and the Chinese will be using some of that iron ore to make bombs to kill Japanese.

Australia has a problem in that in the event of a war with China we now have another lot of citizens with potentially divided loyalties. That is illustrated by how China is conducting its campaign to force some of the wealthy who have fled the Middle Kingdom to bring back the money they left with. A senior fellow at the University of Nottingham's China Policy Institute, Steve Tsang, has said that the Communist Party of China sees itself as wielding dominion over all Chinese people regardless of what passport they may hold. "The party believes if you're of Chinese ancestry then you're Chinese anyway, and if you don't behave like one you're a traitor," he said. Regarding their modus operandi, one Chinese government agent has been quoted as saying," A fugitive is like a flying kite," he said. "Even though he is abroad, the string is held in China. He can always be found through his family." That leads to the possibility that naturalised, Chinese-born Australians may want to be loyal but could be coerced by threats to family members back in China. On top of that, in 2014 there were 94,901 Chinese students in Australia.

While the United States has a number of bases in the first and second island chains east of China, they don't have much strategic depth compared to the force that China will throw at them in a surprise first strike. Thus the US Marine Corps is rotating 2,500 troops through Darwin during the dry season each year. The US Air Force has requested basing B1 bombers at RAAF Base Tindal. The US Navy would benefit from being able to base ships in Exmouth Gulf, as would our Navy.

A scenario for the start of the air war

Chinese military doctrine holds that a war should always be started with a surprise attack because of the advantage it provides. Following is an outline of how China might start its next war. Entitled *Operation Long March*, it was written by Wing Commander Chris Mills (ret.), Royal Australian Air Force.

The Setting: China, a Meeting of the Central Military Commission
The Date: December 6, 2020

The chairman of the Central Military Commission of the People's Republic of China looked at the commission members assembled in the August 1st Building.

The air was charged with tension and expectation.

"Tomorrow, if you each tell me your plans and Forces are ready, we will immediately end the U.S. hegemony of the Western Pacific. The

U.S., despite our repeated warnings, has continued to arm the rebel Government in Taiwan, the latest shipments being 200 F-35s they had surplus after Joint Strike Fighter sales to Europe collapsed. These aircraft are now in action against us; and in the past week, three J-10A 'Vigorous Dragon' fighters have been destroyed while on peaceful patrols of the Straits. This behaviour cannot and will not be tolerated. Now, let me ask about your preparedness—and true and accurate reports only—if there are weaknesses, now is the time to correct them, not in the heat of battle. Commander-in-Chief, tell us the strategic plan."

The CIC rises and opens a PowerPoint briefing on a large screen.

"Our military forces will eliminate the larger U.S. bases across the Western Pacific. Each of the Armed Services will have a substantial role on this joint and highly coordinated operation. The Second Artillery using DF-21 terminally guided Intermediate Range Ballistic Missiles; the People's Liberation Army-Air Force using the new J-20 'Black Eagle' stealth fighters; and, the People's Liberation Army-Navy will employ submarine-launched cruise missiles, mainly our excellent DH-10 missiles. Our Intelligence Agencies have been collecting targeting information on critical infrastructure for several years. The initial targets will be digital communications and military installations and equipment. If U.S. naval forces come within 1,000 nautical miles of our coast, they will be attacked with terminally guided DF-21D missiles. Each of our designated Fire Bases will be protected by mobile surface-to-air missile batteries. Any incoming counter-attack will be detected at long range by high-frequency sky-wave and surface-wave over-the-horizon radars. Preparations will take place in our super-hardened underground airbase hangars to avoid observation by U.S. spy satellites. We have other assets prepared. Mr Chairman, here is the military tasking order for Operation Long March . . ."

There is a sharp intake of breath as the PowerPoint slide flashes onto the screen. The scope and scale of Operation Long March makes the infamous 1941 Pearl Harbour attack look like a mere tactical skirmish.

The members study the PowerPoint slide in complete silence.

The Service chiefs nod their heads in agreement. They know that the deployment of forces laid out in the battle plan is well within their operational capabilities and the skills of their crews.

"A question – why Guam International?" asks the chairman.

"We know that some military aircraft are parked in the hangars to the north of the airfield, but our main target is the fiber-optic hubs near the airport. When we breach those, much of the digital communication across the Pacific to the U.S. will be terminated. This will force the American military and their allies to have to use very limited bandwidth satellites – and we can knock those down too, if we wish," is the reply from CIC.

"Thank you Commander-in-Chief, a good answer. Minister for Finance?" invites the Chairman.

The minister stands and moves to the podium. He speaks confidently but quietly, and the members of the commission strain to hear him. "Our finances are ready to weather the disruption this action may bring. Our GDP is now larger than the whole of the American GDP. We have reduced our exposure in U.S. Treasury holdings to less than $1 billion USD – the Americans' program of printing money has made them virtually worthless as the U.S. dollar continues to decline in value. We have moved our trillions of FOREX USD into the currencies of our trading partners, raising the value of their currency relative to ours so they can afford to buy our goods, and of course making their futures dependent on our holdings. Our gold reserves are over 2,000 tonnes. We plan to sell a large amount when the action starts, while keeping our other precious metal holdings in reserve. We expect the profit will easily offset the cost of Operation Long March. We are in good shape."

Next, the Chief of the People's Liberation Army speaks. "Our DF-21 forces are in position. We plan to launch thirty to sixty rounds into each target."

"Won't that be expensive?" asks the Minister for Finance.

"No", replies the Minister for Defense Materiel. "These missiles are nearly time-expired. It costs as much to refurbish them as to build the new and more effective DF-21Ds."

"Just so" says the PLA chief. "We have upgraded the guidance system to the 'C' model, so they will be very accurate and cost-effective."

As if it's an afterthought, he adds, "I also have the HQ-9, the S-300PMU2, S-400 and the HQ-20 Surface to Air Missiles in place to protect the launch firebases and the key cities of the homeland if the U.S.A. decides to launch cruise missiles from bombers or submarines. We will not get them all, but we will get most and protect our cities and people."

"Air Force?" asks the chairman.

"We have been receiving one J-20 stealth fighter a week since 2016," the PLA Chief continues, "We now have 160 J-20A single-seat fighters and 80 J-20B two-seat theater-bombers. These are all fully operational. You know, U.S. Secretary of Defense Robert Gates was half right when he said China would not have a stealth fighter until 2025 – if he was speaking of our mature and full capability. But I digress. With the low signature of the J-20, we can employ our excellent satellite-guided glide bombs, and the Minister for Finance might be pleased to hear that our estimate of the ordnance cost for an airfield target will be less than $4 million USD. Not a bad price to pay for the destruction of American military capabilities."

"And the Navy?" invites the Chairman.

"We have our submarines in place near each target, and they will be in deep water as they fire, so have an excellent chance of escaping," the Navy chief advises. "This attack will not destroy these bases, but there will be a lot of damage and loss of capability, especially of aircraft the Americans arrogantly park on open tarmacs. I also have underwater assets in place off Los Angeles, and there are several ships carrying containerised Klub cruise missiles off the East and West Coast of the U.S.A., if we need any follow-up action."

"Thank you all. Now, what of the two hundred F-35s on Taiwan?" asks the chairman.

The Minister for Cyber Defense, Madame Chien-Shiung Wu, has a Berkeley Ph.D. in computer science and is young, brilliant, and beautiful – in a scintillating pink-diamond sort of way – and speaks with chilling certainty.

"Over the past decade, my staff have penetrated the F-35 joint strike fighters' software facilities and have made certain changes that cripple them. This is classified 'above top secret,' and you do not need to know more".

The Air Force chief asks, "Then how were my three J-10A's destroyed by the Taiwanese F-35s?"

"A piece of deceptive illusion," replies Madame Wu. "We had to convince the U.S. with a war-like demonstration that their F-35's systems work, so we arranged a successful F-35 attack on the J-10s – when you are network dependent you are cyber-vulnerable. Our pilots were specially briefed and ejected in straight and level flight just before the Taiwanese AIM-120 missiles hit. So the arrogant Americans will think their 'war-tested' F-35s are working perfectly."

"Well, even if they have found this bug," advises the Army chief, "we have been using the F-35 operations over Taiwan to check its signatures, and my SAM people say even the obsolete HQ-9s will kill them. I also have the HQ-20s ready."

"And I have J-10As and Bs and our J-11Bs on alert ready to catch any you don't get," says the Air Force Chief with equally chilling confidence.

"Are you all confident that Operation Long March will be successful?" asks the chairman. All heads nod in agreement. "Very well, the attack on Andersen Air Force Base will start at 10:00 a.m. local time on December 7, 2020. The submarine and DF-21 attacks will be delayed accordingly, as advised by CIC, to provide the Black Eagles with the advantage of complete surprise."

The Air Force chief, General Yónggàn de Zhànshì, rises and asks the chairman, "My son is the J-20B Wing Commander. I am fully qualified on the J-20B. May I fly into battle as one of my son's wingmen?"

"General, you may", agrees the chairman, "and our good fortune flies with you."

The scene moves to Yiwu Airbase near Shanghai, a former People's Liberation Army Naval Air Force super-hardened underground airbase, and now the home of the first J-20B Black Eagle Air Regiment. At 14:00 on December 6. 2020, Colonel Liè Lóng rises to his feet to address his aircrew and battle-staff.

"Warriors, we have been commanded by Air Task Order (ATO) 'Black Eagle Long March' to go into battle to liberate the Pacific Ocean from the seventy-five-year tyranny of the United States. Our regiment's task is to eliminate the Guam Air and Naval Bases as viable military installations. We will . . ." He is interrupted as his airmen rise and begin to cheer – they have been training for this moment for years, and they have great pride in their expertise and equipment, as yet untested in war.

The colonel raises his hand for silence, and the tumult quickly dies away. "As I was saying, we will fly to the Guam launch points--fifty nautical miles for J-20As and 20 nautical miles for the J-20Bs – release our ordnance, and return. The intelligence and weapons officers have done their jobs well, and each bomb has been programmed with a target area. Some of the 100-kilogram bombs with electro-optical sensors will search for-high value targets, and other 500-kilogram bombs will use Global Positioning System-Inertial Navigation System to destroy fixed installations. We have three squadrons of twelve J-20As and a squadron of twelve J-20B fighter-bombers for the task, so this will be a near-saturation attack."

"J-20B pilots, you will each be flying the Deadly Rain manoeuvre you have practised so many times. Make sure you fly with precision and confidence, as your lives and the success of your missions will depend on your skill. The J-20As will also be armed as fighter escorts in case any F-22As or F-35As are on combat air patrol duties. Study your individual orders carefully. Now, get some rest. We take off at 06:30, there is a refuelling outbound, and possibly inbound, if you are intercepted. The mission will last about seven hours. China salutes your strength and courage."

General Yónggàn de Zhànshì rises at 04:00, dresses and enjoys a hearty traditional breakfast, and attends the final briefing at 05:30. Weather is clear over Guam, as the 'fanumnangan' dry season has started – a perfect environment for smart electro-optical bombs looking for exposed targets. At 6:30 he is taxiing as number two in the third element of the second squadron and makes a smooth uneventful take-off. Captain Gōngjiàn Shou, his weapon systems officer, is nervous flying a combat mission with the chief at first, but he soon settles into the task.

The squadrons fly out one thousand nautical miles, and take it in turns to drink from an H-6U Badger and an Il-78 Midas aerial tanker. Then on another 600 nautical miles to the weapons-release point.

Like most war activities, this long straight and level transit flight is boring. Each J-20's radio frequency surveillance system is active, however, continually 'sniffing' the ether for hostile radar and radio transmissions. As the planes approach Guam, American radars and civil aircraft chatter are intercepted and assessed. Each of the J-20 fighters is part of a low-probability-of-intercept information net, with data being exchanged by directional millimetre-wave data-link pencil beams. The network has multiple redundancies, and each and every aircraft can act as a peer-to-peer node. As a result the aircraft all share a common air picture, and the crews can communicate with little chance of the transmissions being intercepted.

A professional calm prevails in the J-20s as each of the aircraft approaches the GPS release point designated in its orders. As the J-20Bs are about twenty nautical miles from Guam at 45,000 feet and Mach 1.5, the engines are advanced to full afterburner, and the noses raised to 20 degrees; 0.8G is held.

At the designated release point, the weapons bay doors on each aircraft are opened and the LS-6 100 kg smart bombs are ripple-released at one-per-second. Lastly, each plane rolls a few degrees to the right and launches a single LD-20 decoy dispenser. Doors are closed and the aircraft barrel-roll into a tight Immelman turn to escape, presenting any missiles that may be launched from Guam with a very difficult supersonic tail-chase. Engines are returned to military power to close the nozzles and lower the radar signature.

Meanwhile, the two hundred and forty 100-kilogram LS-6 smart bombs fall into the attack "basket," descending in a graceful curve.

Each LD-20 decoy-dispenser falls five thousand feet and its three petals open, releasing 49 radar-reflecting decoys, each with aerodynamic drag designed to fall slightly faster than the LS-6 bombs. This, plus the chaff packed throughout the canister, screens both the attacking LS-6 bombs and the retreating J-20B aircraft.

At 09:56 on Guam, it is a beautiful clear Monday morning. The MIM-104 PAC-4 Patriot teams are recovering from a weekend of hectic social activities and enjoying the light sea breeze. Their reverie is rudely broken into by a blaring klaxon. Lieutenant Michael Brown, in charge of the battery, bounds into the Engagement Control Station van.

What he sees first confounds, and then horrifies him – forty-eight radar symbols bloom across the screen, and then from each symbol a cloud expands with symbols too many to count – all inbound for Guam. Some of them are headed directly for Andersen Airforce Base, others for the Guam International Airport, and a third set to the Apra Naval Complex. A gut-wrenching comprehension of what is unfolding dawns on the lieutenant.

"This is a stealth attack – and those incomings are probably bombs." Forty-eight of the symbols disappear as the J-20 weapons bay doors close. A faint, rapidly retreating group of blips appears; then they wink out one by one off the screen.

The larger 500-kilogram winged LS-6 glide bombs head for their GPS-designated targets: C3 centres, maintenance facilities, munitions storage, and the massive underground fuel storage tanks.

The smaller 100-kilogram LS-6 small diameter bombs have a more interesting task. As they approach their GPS-designated search box, their electro-optical seekers scan the tarmac, selecting high-value targets such as B-2As and F-22As over lesser-value ones like F-35s and F/A-18s.

They do not lock in their aim-point selection until the final few seconds of flight.

If they don't find a parked aircraft target, they head for a building. Some 240 of these LS-6s are incoming, screened by no less than 588 decoys, which have nearly identical radar signatures.

Lieutenant Brown is not having a good day. He finally orders the battery "weapons free" to engage as many incoming bombs as possible. His Patriot launchers have up to forty-eight ready shots loaded, but with 240 bombs and 588 decoys incoming, the task he is faced with is impossible. The Patriot missiles scan ahead and each detects and reports a plethora of returns using its track-via-missile system. But which is a bomb and which is a decoy? After all rounds are fired, more than two hundred 220 LS-6s are still inbound, and nothing is left in the Patriot locker to fire.

Two B-2As and twelve F-22As are on a deployment to Andersen AFB. They are just back from the 06:00 'sunrise strike' on Farallon de Medinilla Island bombing range and are being refuelled, repaired, and re-armed for their next training mission at 12:00. Without hardened shelters, each aircraft is in the open, and several of the LS-6s find them in their designated kill box.

The LS-6s arrive nearly simultaneously like a deadly hail from Hell, and the entire tarmac area erupts in a massive series of explosions, enhanced by aircraft fuel, tankers, and weapons sympathetically exploding. The ground crew watch, horrified, as the LS-6 smart bombs drop nearly vertically into the centre of each aircraft, blasting the planes to smithereens.

At Guam International Airport, things are much the same. PLA HUMINT (People's Liberation Army Human Intelligence) has identified which of the hangars on the north-side of the airfield contain military aircraft, and several are hit by 500-kilogram LS-6 glide bombs. More importantly, the communications buildings across Guam housing the trans-pacific fiber-optic cable repeaters are hit with several bombs, and communications are instantly terminated.

Figure 29: Johnson South Reef, 27th May 2015 (image courtesy of Victor Robert Lee)
China has dredged channels into some of the reefs and used the spoil to build islands. At the southern end of this island on Johnson South Reef is one of the forts under construction with at least two hexagonal flak towers. The forts all have ramps up to the first floor.

The Apra Harbour Naval Complex receives multiple hits from the assigned 72 LS-6 500-kilogram GPS-INS guided bombs. HUMINT delivered by cellular telephone earlier that morning has identified two nuclear submarines and three frigates alongside piers. Each receives a direct hit by an LS-6. The remaining rounds devastate the support facilities.

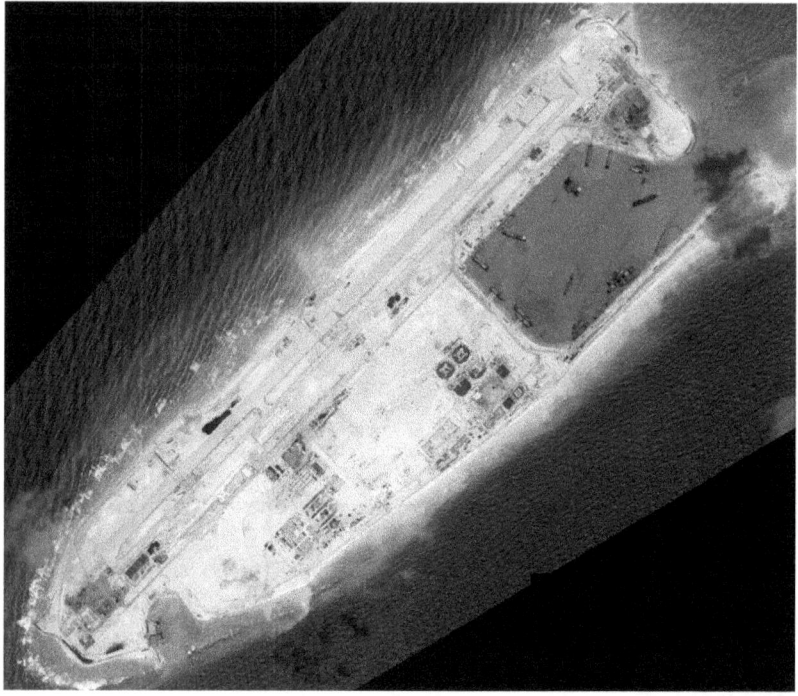

Figure 30: Fiery Cross Reef, 13th July 2015 (image courtesy of Victor Robert Lee)
The airstrip and adjacent taxiway on Fiery Cross Reef are both three kilometres long. This airfield, 1,000 kilometres from Hainan Island, will be able to land all the types of aircraft in the Peoples Liberation Army inventory and thus extend the range of Chinese aircraft a further 1,000 kilometres from the Chinese mainland.

Ten minutes later, a pair of J-20R reconnaissance fighters flying from the southeast pass over Guam, with AESA radars and cameras recording the damage. Their job is hampered by massive quantities of burning fuel and aluminium ash in the air, but the nevertheless AESAs can detect the detail of shapes on the ground.

Several hours later, General Yónggàn de Zhànshì is enjoying the euphoria with the aircrew and support staff in the Black Eagle Operations Room. "I cannot, for security reasons, tell you more, but China is very

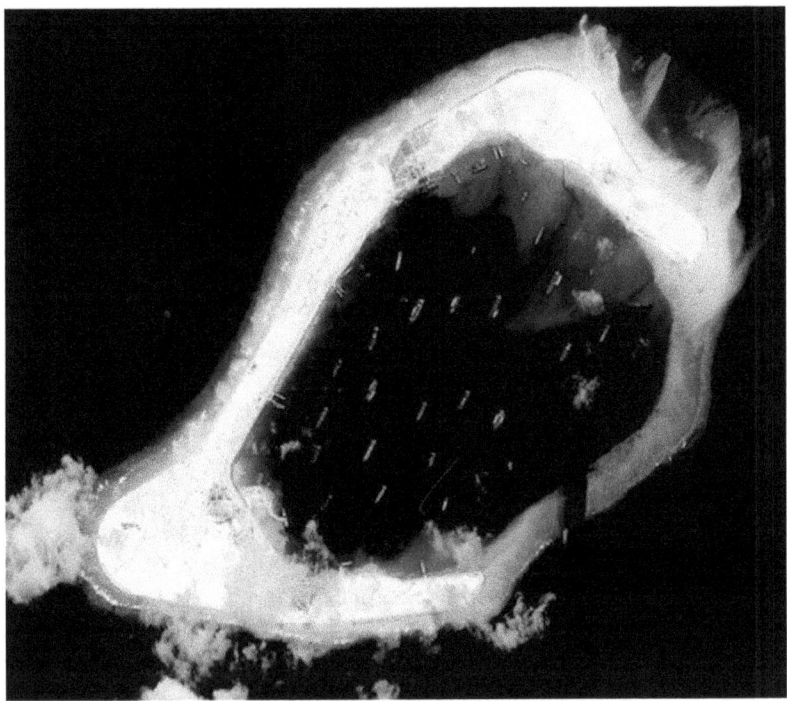

Figure 31: Subi Reef, 13th July 2015 (image courtesy of Victor Robert Lee)

There are about 50 vessels in the lagoon of Subi Reef in this image, now a harbour fomed by dredging the lagoon and piling the spoil on the reef. The straight section on the western side is three kilometres long and is being compacted to prepare it for an airfield. The lagoon is 3.5 kilometres long by 2.4 kilometres wide. Mischief Reef to the east of Subi Reef is another atoll that is being converted to take an airfield. The lagoon at Mischief Reef is eight kilometres long and will be able to shelter a substantial number of ships.

proud of you, and those who designed and delivered the Black Eagle capability."

He then returns to his J-20B and, with Captain Gōngjiàn Shou, flies to Beijing International, where the J-20B stealth fighter will be put on public display. From there, he is whisked away by a staff car to the Central Military Commission.

The Chairman of the Central Military Commission addresses the assembled members. "Well done to each of you and the Services you represent. In Operation Long March the time from first J-20 weapons-bay door opening to the last DF-21 and DF-10 impact was less than fifteen minutes. Twenty strategic targets were attacked and severely damaged across 6 million square miles of ocean, and we had no losses. Assessments are still being conducted, but it seems that all strategic objectives have been achieved. We must now be on our guard for a counter-attack. We are ready. Thank you all."

Conclusion

China is preparing for war, not just being better able to defend itself. They are most probably thinking of a quick amphibious operation to scrape all the other countries in the South China Sea off their islands, followed by a another amphibious operation to sieze the Ryukyu Islands from Japan. Appendices three to six present other views on China's irredentism from experienced observers.

6

GOING NUCLEAR

Introduction

The Cold War really was a benign period in history. Modern civilisation had just come out of a World War which had slaughtered civilians at an average rate of 23,000 per day. The nuclear rivals confronting each other in the Cold War had no stomach to repeat such mass slaughter that was fresh in their minds. They thus fought minor wars and skirmishes by proxy but otherwise got on with the business of either advancing their civilisations or stagnating. None of the parties with nuclear weapons during those decades seriously considered using them since the consequence would be their own annihilation.

Eventually the Soviet Union and its empire became exhausted and the ideological basis of communism was discredited by just how backward that country remained. That was a good thing. One consequence of communism's collapse is that the former communists and their fellow travellers in the West found a new ideological home in the environmental movement and moved to promoting the global warming hoax as a wealth redistribution exercise via a network of United Nations agencies. Another consequence is that a number of errant regimes felt much less constraint in developing nuclear weapons.

The errant nuclear power of the moment is Iran, which has a large, well-funded uranium enrichment program and a stated intention to annihilate Israel with nuclear weapons. Whatever the fate of the Iranian bomb-making effort, there is another state which is heading towards failed state status while still upping the bomb-making rate of its nuclear weapons program. This is Pakistan, land of the pure. This is in one of the world's poorer countries with a literacy rate of 55 per cent and a population growth rate of 1.7 per cent per annum. Yet it is believed to have an arsenal of approximately 100 completed nuclear weapons and is accelerating its bomb-making program. Pakistan is a failed state in waiting. When it does fail, what will be the fate of all those nuclear bombs? This situation is not going to end well.

Nuclear weapons primer

The fact that uranium could be split with a neutron was discovered by Otto Hahn and Fritz Strassman in Berlin in late 1938. The nuclear physics community promptly realised that this could become the basis for a bomb. Einstein's letter of 2nd August 1939 to President Franklyn Roosevelt warned of the potential danger of a German nuclear weapons program. There is one naturally occurring isotope of uranium, U^{235}, that can be used for a fission weapon. To make a weapon using U^{235}, it needs to be enriched from its level in uranium ore of 0.7 per cent to beyond 80 per cent This is an energy intensive process relying upon the slight difference in mass between U^{235} and U^{238} atoms. The only artificial element practical for a fission weapon is a plutonium isotope, Pu^{239}. To make Pu^{239}, U^{238} is irradiated in a reactor with neutrons.

Pu^{239} is marginally cheaper to make than U^{235}. Its one big advantage over U^{235} in bomb-making is that its fission cross-section is four times larger, which means that a Pu^{239} atom is four times more likely to be split by a neutron that hits it than a U^{235} atom. In turn, this means that the fissile core of a Pu^{239} weapon can be one quarter of the weight of U^{235} weapon.

However, Pu^{239} comes with some significant drawbacks. The first plutonium isotope made in a reactor is Pu^{239}. If an atom of Pu^{239} is hit with a neutron, there is a 62 per cent chance that it will split and a 38 per cent chance that it will accept that neutron and become Pu^{240}. In a nuclear reactor operated for power generation, this process continues on such that higher isotopes of plutonium are also created. When fuel rods are extracted after their normal three year life span, the mix of plutonium isotopes present is 53 per cent Pu^{239}, 25 per cent Pu^{240}, 15 per cent Pu^{241}, 5 per cent Pu^{242} and 2 per cent Pu^{238}. Only the odd-numbered isotopes are fissile though. By the time fuel rods are pulled, fission of plutonium is providing half of the energy created. To mitigate the problem of neutron capture by Pu^{239} in making weapons grade plutonium, the time that U^{238} is kept in the reactor is limited to a few weeks. The production rate of a reactor specifically designed for plutonium production is one gram per day per megawatt of thermal capacity. Consequently, for a 100 megawatt reactor, this is equivalent to the weight of 11 one dollar coins per day.

The problem with Pu^{240} is that it has a high rate of spontaneous fission with the result that plutonium with a high proportion of Pu^{240} tends to detonate prematurely. That, in turn, causes a low yield. Weapons-grade plutonium is made with no more than seven per cent Pu^{240}. Plutonium cores also generate a considerable amount of heat from isotopic decay. Depending upon size, the core of a plutonium weapon may be at 200°C. This affects the high explosives surrounding it. Plutonium weapons need expensive ongoing maintenance to reprocess the cores and replace the chemical explosive. While Pu^{239} has a half-life of 24,000 years, U^{235} has a half-life of 704 million years and thus a much lower rate of spontaneous fission. There has been a swing towards U^{235} cores in the US nuclear weapon inventory due to this resultant maintenance question.

Scientists working in the Manhattan Project during World War 2 realised they had two isotopes to work with, U^{235} and Pu^{239}, and two possible designs – the gun type weapon, and the implosion of a sphere. Only U^{235} can be used in a gun-type weapon due to the Pu^{240} problem causing premature detonation in this design. The main problem with the implosion design is getting the high explosives surrounding the core to compress it evenly and keep it spherical as it is compressed to criticality. The Manhattan Project's scientists were so certain that the gun-type weapon would work that they did not bother to test it before using it in the war. They were less certain about the implosion design, so the first nuclear test, codenamed Trinity, conducted on July 16, 1945 was of a plutonium bomb. The next nuclear explosion was the gun-type U^{235} weapon dropped on Hiroshima. This was followed by the implosion Pu^{239} device dropped on Nagasaki. The gun-type weapon using U^{235} is considered to be a low tech weapon that would not be difficult to make. It would also be reliable. The downside is the low yield relative to the amount of fissionable material used and this in turn means a much heavier bomb. For example, the Nagasaki weapon used 6.2 kg of plutonium of which about 1 kg (17 percent) fissioned to produce a yield of 21 kilotons. By comparison, the Hiroshima weapon used 65 kg of 80 per cent U^{235} of which less than one kilogram fissioned for a yield of 13 kilotons.

Countries developing nuclear weapons today tend to forego the gun-type design and use implosion designs only. The exception was South Africa which built five gun-type bombs during the 1980s and then dismantled them at the end of its apartheid period. South Africa's design used 55 kg of 90 percent-enriched U^{235} with an estimated yield of 10 to 18 kilotons, which is an efficiency of 1.0 to 1.8 percent. The cost of the South African nuclear weapons program is estimated to have been in the order of $240 to $310 million, showing just how affordable a nuclear weapons program can be.

Fission weapons can be boosted for a dramatic increase in yield. In an un-boosted plutonium weapon, a maximum of 20 per cent of the core will fission before it blows itself apart. Boosting involves injecting a mixture of deuterium and tritium (isotopes of hydrogen with one and two neutrons respectively) into the core just prior to detonation. By the time one per cent of the core has fissioned, the temperature has risen high enough for the deuterium and tritium to fuse, releasing a large amount of high energy neutrons. The result is that the amount of the core fissioned will approach 50 per cent with a corresponding increase in yield. The tritium itself, typically 3 grams, contributes less than two per cent of the yield. The problem with tritium is that it has a 12.4 year half-life so that the tritium bottles in nuclear weapons keep on having to be refreshed.

The next step up from fusion-boosted weapons is fusion weapons in which a fission primary initiates an attached fusion secondary – the hydrogen bomb. In theory, there is no upper limit on the potential size of a fusion bomb. The largest fusion device tested was the Russian "Tsar Bomba" of 50 megatons in 1961. The Hungarian- American physicist

Edward Teller calculated that anything larger than 100 megatons would simply blow a segment of the Earth's atmosphere out into space. The most modern US warhead, the W88, has an estimated yield of 475 kilotons and is thought to weigh less than 360 kg. Most of the yield of hydrogen bombs is from fission of the U^{238} mantle by high energy neutrons from the fusion reaction. Neutron bombs are fusion devices without the U^{238} mantle.

Nuclear Weapons in South Asia

Pakistan's nuclear history is intertwined with India's. Their nuclear weapons programs responded to each other's efforts and so it is appropriate to consider them together. India began with a proposal for a nuclear research institute as long ago as in 1944, a year before the atomic bomb was dropped on Hiroshima. By 1955 India's effort moved in the direction of weapons manufacture with the acquisition of a 40 megawatt heavy water reactor from Canada. This reactor was ideal for manufacturing weapons grade plutonium and it was duly used to produce up to 40 grams daily of weapons grade plutonium – enough for a couple of weapons a year. However, the Indian program encountered trouble with their plutonium reprocessing plant and it took until 1969 before enough plutonium had been separated to make one bomb.

China attacked India in 1962 and seized the Aksai Chin plateau in Kashmir. The first Chinese nuclear test was two years later and the combination prompted India to speed up its nuclear weapons efforts. In turn, Pakistan became agitated by the Indian nuclear effort. Then Foreign Minister, later President, Zulfikar Ali Bhutto stated in 1965 that, "If India builds the bomb, we will eat grass or leaves, even go hungry, but we will get one of our own. We have no other choice." Pakistan started a one month war with India that year. India and Pakistan fought again in 1971, the war that resulted in East Pakistan becoming Bangladesh. This latter war is very telling about the Pakistani character. The war was triggered by an East Pakistani party winning the election that year. Rather than handing over control of the country democratically, the Pakistani Army decided to slaughter East Pakistanis and to decapitate their society.

The head of the Pakistani Army, Lieutenant-General Gul Hassan Khan, declared,"Kill three million of them, the rest will eat out of our hands." In what was called Operation Searchlight, the Pakistani Army duly killed three million people in what is now Bangladesh and in particular sought out and killed the Bengali intellectual, cultural and political elite. In that respect, it combined elements of the Nazi holocaust in Poland with Russia's massacre of Poland's elite at Katyn Wood.

The Indian Prime Minister, Indira Gandhi, gave the go-ahead for building a nuclear weapon in September, 1972. The first Indian nuclear test was conducted eighteen months later on 18th May 1974. It was a solid core device using 6 kg of plutonium and had a yield estimated at 8 kilotons. By 1983 the Indian nuclear establishment was ready to test two new designs. Though not receiving approval for this move from the Indian Government, the country nevertheless launched its missile program intended to deliver nuclear warheads. This resulted in the development of the Prithvi with a range of 150 km and the Agni with a range of more than 1,500 km. In 1985 Indian defence planners were envisaging an arsenal of 70 to 100 warheads for a total outlay, including delivery systems, of $5.6 billion. The move to integrate nuclear weapons with delivery systems began in 1986.

President Bhutto of Pakistan also decided to proceed to a nuclear weapon in 1972. Two years later he arranged for Libyan and Saudi funding of the program. At the same time, Dr A.Q. Khan, a Pakistani national who had worked with centrifuge design in the Netherlands, had been able to smuggle centrifuge blueprints into Pakistan. Enrichment of U^{235} commenced in 1979. The Soviet Union's 1979 invasion of Afghanistan would provide the Pakistanis with a lot of political leeway with respect to its relations with the United States. Then in 1983 China would provide Pakistan with the design blueprints for its fourth nuclear test - a solid core device that yielded 25 kilotons. In that same year, Pakistan achieved 90 per cent enrichment of U^{235}. Its enrichment facilities are based in the town of Kahuta, just 30 km southeast of its capital, Islamabad.

Pakistan's first cold test (without a fissile core) of a weapons design was in 1984. And it started stockpiling completed weapons in 1989. Prime

Minister, Benazir Bhutto, Zulfikar Bhutto's daughter, was updated on her country's weapons progress by US officials since Pakistan's military would not tell her what its scientists were doing. In 1990 the United States estimated that Pakistan had made 125 kg of weapons-grade U^{235} and three years later Pakistan sought and received North Korean missile technology The annual budget of the Pakistani nuclear program was $20 to $25 million with the total program cost over 25 years of less than half a billion dollars.

After he had finished arming Pakistan with nuclear weapons, Dr Kahn resigned from the civil service and set up in the business of selling nuclear weapons technology to several other Islamic countries. Algeria, Egypt, Syria and Saudi Arabia declined his services. However, he was able to do business with Iran and Libya and set up a smuggling network to those countries and possibly to North Korea.

India had a combat-ready system in 1994, which was demonstrated by a Mirage 2000 dropping a complete bomb, minus its plutonium core. On May 11, 1998, it tested three devices and followed this up with a test of two sub-kiloton devices two days later. In response to these Indian tests, Pakistan conducted tests on May 28 and 30, with yields of 9 kilotons and 5 kilotons respectively. Pakistan's nuclear sector made a lot of other progress that year. The first plutonium-producing reactor at Khushab became operational and its first test of a long range missile was conducted. This was a North Korean-supplied No-dong missile. And Pakistan attacked India again in Kashmir, in 1999, beginning what came to be known as the Kargil war, which cost Pakistan 474 lives.

The country's pace of activity, both in nuclear weapons expansion and attacks on India, did not, however, slow down into the next century. In 2001 a terrorist attack on the Indian Parliament was launched from Pakistan. Construction began on a second plutonium-producing reactor at Khushab in 2002 and a third in 2006. A Pakistani initiated attack on India's coastal city of Mumbai with terrorists backed by Pakistan's secret service agency, the ISI, would follow in 2008. The terrorists were members of the Lashkar-e-Taiba, the same group that had earlier attacked India's Parliament. In 2011satellite imagery showed that construction had commenced on a fourth plutonium-producing reactor at Khushab.

Pakistan's uranium enrichment facilities are thought to be capable of producing 110 kg of weapons grade U^{235} annually, which is enough for five weapons. On the completion of the fourth Khushab reactor, and assuming that these four reactors are 70 megawatts thermal units and have 70 per cent availability, Pakistan could produce 70 kg of weapons grade plutonium each year, enough for fourteen weapons. At the combined rate of possibly 19 weapons annually, at the end of this decade Pakistan would be one of the world's larger nuclear powers with an arsenal equivalent in size to those of the United Kingdom and France. It is likely that the increase in build rate is driven by the needs of the program's financier, Saudi Arabia. The Saudis' major concern is Iran's ongoing acquisition of nuclear weapons, a policy that is being driven by a desire to emerge as the Persian Gulf region's hegemon unless countered by an equal or greater nuclear arsenal. Just as the prospect of India acquiring nuclear weapons initially drove the Pakistani program, Saudi Arabia's ruling family has opted for the same path. Saudi King Abdullah said in 2009, with reference to Iran," If they get nuclear weapons, we will get nuclear weapons."

Indeed, Saudi Arabia may have already taken delivery of Pakistani-manufactured nuclear warheads. It may have done so to prove that the acquisition system works. It is quite likely that the accelerated pace of Pakistan's bomb-building was in fact to meet a Saudi order. Just how many bombs the Saudis think they require is indicated by the size of the Saudi ballistic missile fleet. In the late 1980s, the Saudis purchased DF-3A single stage, liquid fuelled missiles from China. The number acquired may range between 30 missiles and nine launchers to 120 missiles and 12 launchers. Its missile bases are situated at Al Joffer and Al Sulayyil, about 90 km and 450 km southwest of the capital, Riyadh, respectively. Continued Chinese presence at the bases is required for technical support, maintenance and training. Saudi Arabia is currently seeking a solid fuel missile to replace the inaccurate liquid-fuelled DF-3A. This replacement may well be China's DF-21 or a Pakistani version of this weapon. Ballistic missiles are an expensive way of delivering high explosives. There is no doubt that the Saudi missile fleet was created with the intent of mating it with nuclear warheads.

Saudi Arabia's little-known rush to a more capable nuclear force would have been given a push-along by the Obama Administration's abandonment of the Mubarak regime in Egypt during the so-called Arab Spring of 2011 and 2012. The Saudis would have realised that they are now on their own. The weapons build to fill the Saudi order may well take the Pakistanis the rest of the decade to produce and assemble.

Interestingly, just as nuclear weapons have been coming to the fore across the Middle East, the ability to wage conventional war has declined dramatically. The reason is that all the players now import a high proportion of their food requirements. For example, the last time Egypt and Syria attacked Israel was in the Yom Kippur War of 1973. Then they had populations of 38 million and 8 million respectively. Today their population levels have risen to 82 million and 21 million respectively and all of that increased population is being fed with imported food. It is hard to project power conventionally when your own population is on the edge of serious food shortages, and thus likely starvation. This subject is covered in the following chapter.

Failed state is baked in the cake

While Pakistan remains a highly dysfunctional society, it is a significant grain exporter. Wheat production rose from 4 million tonnes in 1960 to 24 million tonnes in 2011. That is an increase of 500 per cent in fifty years. However, wheat production per capita has been flat at about 140 kg since 1980 as population has increased in tandem. Pakistan currently exports 4 million tonnes of rice annually. Let us assume that an adult Pakistani can be adequately fed with 300 kg of rice a year. That 4 million tonnes will therefore feed just over 13 million Pakistani adults. Presently, Pakistanis are being created at the rate of 4.6 million annually and are dying at a lower rate of 1.3 million per annum. This net increase of 3.3 million per annum will account for the current surplus of exported rice within another five years or so when the population will reach 200 million.

Pakistan has been described as a "militarist-Islamist nation". But it boasts a literacy rate of only 55 percent, and its cities and towns experience frequent power blackouts. When will it all end? The end of the present decade may well be accounting time. A synchronous decline in world grain production due to solar-driven cooling of the climate would mean that grain would be unavailable at any price. A society tipping over into starvation will eat its seed grain, causing complete collapse. The individual nuclear weapons in the Pakistani arsenal would become saleable items for the military commanders who could get hold of them.

Between 2002 and 2010 the United States provided Pakistan with some $18 billion in military and economic aid, and a further $3 billion is in train. As such, the United States has allowed Pakistan to acquire the financial capacity to fund its nuclear weapons program and thus the creation of nuclear weapons that many Pakistanis would wish to use against the United States. In July 2012, a Pew poll found that 74 per cent of Pakistanis consider the United States to be an enemy.

Pakistan's inherent barbarism was recently demonstrated by an attack using its Special Services Group (SSG) commandos on Indian soldiers guarding the Kashmir border with Pakistan on 8th January 2013. The incident resulted in two Indian soldiers being beheaded. The SSG is Pakistan's main special forces unit and operates under direct orders from the military high command.

Israel

Israel's heavy water reactor for making plutonium went critical in 1964. It is thought to be sized at 150 MW thermal which could produce 40 kgs of plutonium annually. By using 4kg per weapon, that is adequate for 10 warheads each year. Israel may therefore now have in the order of 400 warheads. Most of these are thought to have yields of 200 kilotons, which means they would have a tritium boosted primary stage and a fusion secondary stage. A yield of 200 kilotons appears to be the optimum trade-off between weight of the bomb and blast area. Delivery systems include fighter aircraft such as F15s, missiles and submarine-launched cruise missiles with a range of 1,500 kilometres. Israel keeps a submarine on station south of Iran in the Indian Ocean to ensure a second strike capability.

Algeria

The United States-led 2003 invasion of Iraq flushed out Libya as a state that was in the process of developing nuclear weapons. Libya promptly voluntarily gave up its uranium enrichment program. A bit further to the west, Algeria retains its nuclear weapons effort. Algeria has a 15 MW thermal heavy water reactor in the north of the country, which could be used to make plutonium. Its secret nuclear weapons program is located south of the town of Tamanrasset near the southern border with Mali. This initially complicated efforts to engage elements of Al Qaeda that had established a base in northern Mali, centred on the ancient city of Timbuktu. The Algerian Government was afraid that foreign military staging through Tamanrasset might have taken the opportunity to destroy its nuclear weapons work. Significantly, Algeria bases some of its Su-30 fighter aircraft at Tamanrasset.

Iran

Iran is in the process of building its own nuclear weapons capability. It has already the missile delivery systems to mate with those weapons. It may also already have a small number of weapons smuggled from the Soviet stockpile during the collapse of the Soviet Union in the early 1990s. The question regarding Iran is intention. Will the Iranian Government use its future nuclear weapons inventory to enhance its position as a regional hegemon or will it use them in a first strike attack on Israel as it has repeatedly stated that it will?

The answer is yes: it will use those weapons to attack Israel even though it knows the result will be the destruction of the Islamic Republic of Iran. The clerical elite that controls Iran considers the country as being more of a process than a state. The desired outcome of that process is conversion of the entire world to Islam. Iran is a Shi'ite state in which many of the clerical elite believe in the Twelfth Imam, also known as the Hidden Imam. The Twelfth Imam, called the Mahdi, is an historical figure born in 869 AD and who disappeared in 941 AD. His disappearance is referred to as the Occultation. Twelver Shi'ites believe that the Mahdi will appear with Jesus Christ as his sidekick to bring justice to the world. They also contend that they should hasten the return of the Mahdi by creating conditions that bring this about. This mainly involves slaughtering non-Moslems or their forced conversion.

Iran has been conducting low-level warfare against the United States and other western states since the Islamic revolution in 1979. This warfare has mostly taken the form of state-sponsored terrorism. In fact Iran hosts an internationalist terrorist confab in Tehran each February. State-sponsored terrorism should be considered as a form of war and treated accordingly. The head of Iran's Council of the Guardians of the Constitution, Ayatollah Ahmad Janatti said on Iranian television in 2007 "Just like this movement destroyed the monarchical regime here, it will definitely destroy the arrogant rule of hegemony of America, Israel, and their allies. At the end of the day, we are an anti-American regime. America is our enemy, and we are the enemies of America. The hostility between us is not a personal matter. It is a matter of principle. We are in disagreement over the very principles that underlie our revolution and our Islam." The United States may not be interested in Iran, but Iran is very much interested in the United States.

Herman Kahn's book on the theory of how to conduct nuclear warfare, *On Thermonuclear War*, is over 600 pages long. Iran's philosophy on conducting the nuclear war it wishes to have may be summarised in a few sentences by a former president of the republic, Ayatollah Hashemi Rafsanjani who in December 2001 said," the use of even one nuclear bomb inside Israel would destroy everything" "while the Muslim world, if attacked in retaliation, could easily afford to lose millions of "martyrs".[1] He concluded: "It is not illogical to contemplate such an eventuality." There is plenty of evidence that the personal beliefs of the Iranian clerics match their public statements.

Based on over thirty years of statements and actions, it is a question of when and how Iran will attack Israel and possibly the United States. What weapons does Iran have to realise this wish? Iran has a broadly based and well-funded nuclear weapons program which includes multiple sites for centrifugal enrichment of uranium, a heavy water plant sized at 100 tonnes annually, and a 40 MW thermal heavy water reactor suitable for production of plutonium. By 2003 Iran had created an isotope of polonium (Po^{210}) by irradiating bismuth in its Tehran reactor. One of the best-known uses for this isotope is as a neutron initiator in nuclear weapons.

It has been estimated that as at November, 2012 Iran had 7.6 tonnes of uranium enriched to 3.5 per cent U^{235} and 232 kg of uranium enriched to 20 per cent U^{235}. Further enriched, this is estimated to provide enough uranium to make five weapons with cores of 16 kg of 90 per cent U^{235}. Iran has developed a range of short range ballistic missiles to mate with future warheads.

Guaranteed second strike

The US-Russian nuclear stand-off gave mankind the most peaceful period in history. This was the period of what would be dubbed mutually assured destruction, or MAD, which made the nuclear powers cautious in their dealings with each other. We no longer have MAD to keep the peace and the international fabric of humanity is being degraded by questions such as whether or not Iran will launch an "out of the blue" nuclear strike on Israel as soon as it is capable of doing so.

Professor Paul Bracken in his book "The Second Nuclear Age"[2] recommends that the United States keeps the peace with a policy of "No first use – guaranteed second use". As he has noted, the United States, while not ever having renounced first use, has had a *de facto* policy of "no first use". The important point is the second half of Professor Bracken proposal – "guaranteed second use". This means that the United States will retaliate against any country's first use of nuclear weapons to attack any other country. The consequence of this is that any non-nuclear power has the deterrence effect of the 4,000-odd warheads of the United States arsenal. His proposal might be enhanced by further detail, which is that as part of its "guaranteed second use", the United States would conduct a decapitation strike on the country that first resorts to nuclear weapons use outside its own borders and that that country's capital would be destroyed. "Guaranteed second use" is an appropriate evolution of the Truman Doctrine for our age of nuclear proliferation. As said by President Truman, his doctrine was "the policy of the United States is to support free peoples who are resisting attempted subjugation by armed minorities or outside pressures". A further enhancement would be to offer Russia a role in enforcement.

The alternative is that nuclear proliferation will ripple out from Iran. For example, Azerbaijan offered the use of its airbases for Israeli aircraft attacking Iran. The Azerbaijanis are well aware that Iran will use its coming nuclear weapon status to bully all its non-nuclear neighbours.

Also, if nations commenced nuclear exchanges with 20 kiloton weapons, they will soon find that just a few of these will not have much of an effect. Study of the flat terrain around Hiroshima found that the zone of destruction was defined by the area of 5 psi or greater overpressure. For a 20 kiloton warhead, that means that the area destroyed is 12.6 sq km. Let's assume that Iran gets a 20 kiloton warhead past Israel's ABM system and it landed on Tel Aviv. That coastal city has an average population density of 7,300 people/sq km so the Iranian warhead will kill just over 90,000 people which is three per cent of the population of the city. True, that would be tragic. But Israel would survive without much impairment of its war-fighting capability. Unless there is massive retaliation for the first nuclear strike, wherever it is, the genie will be out of the bottle. Nations will find that they can survive the exchange of a few weapons and will plan on using them. Mankind's existence on planet Earth will become much, much darker.

Living with fallout

As the number of nuclear weapon states on the planet increases, so does the chance that one of these states will seek to settle its differences with its neighbours with nuclear weapons. Similarly, the natural proclivity of uranium-burning light water reactors is to explode unless cooling water is kept up to them. So as more of this type of reactor is built, the chance of a Fukushima-type incident also increases. The use of nuclear weapons used against countries with uranium-burning light water reactors for power generation will be quite synergistic in terms of the quantum of radioactive fallout produced. That might for as simple a reason as the backup-cooling systems running out of diesel before the reactor core reaches cold shutdown, which could take months.

With the potential for having to live with fallout increasing, it is apposite to see what the effects of fallout are, and how to remedy contamination. There are two well-studied examples to draw conclusions from: the nuclear attacks in Japan in August, 1945 and the Chernobyl incident. The Hiroshima and Nagasaki bombs were both detonated at an altitude of approximately 1,600 feet. The former was a 16 kiloton uranium bomb and the latter a 21 kiloton plutonium bomb. At Hiroshima buildings over 10 square kilometres of the city were destroyed with about 60,000 dying immediately from blast, thermal effects and fire. Within four months the death toll rose to about 150,000. At Nagasaki immediate deaths may have been of the order of 40,000 with total deaths rising to perhaps 80,000 within four months. A group of 87,000 survivors of both bombings who had been exposed to radiation were followed in health studies over 60 years In that group of 87,000, there were 430 more cancer deaths than would be expected in a similar but unexposed population (8,000 cancers from all causes compared to an expected 7,600). This is an increase of 0.5 per cent in the total population. These bombings were airbursts which produce much less fallout than ground-bursts.

The now-closed Chernobyl reactor complex is 100 km north of Kiev in the Ukraine. The Unit 4 reactor was due to be shutdown for routine maintenance on April 25, 1986. It was decided to undertake a test of the capability of the plant equipment to provide enough electrical power to operate the reactor core cooling system and emergency equipment during the transition period between a loss of main station electrical power supply and the start-up of the emergency power supply provided by diesel engines. Unfortunately this test, which was considered to concern essentially the non-nuclear part of the power plant, suffered from a lack of coordination between the team in charge of the test and the personnel in charge of the operation and safety of the nuclear reactor. The test operators deviated from established safety procedures. Their actions were compounded by significant drawbacks in the reactor design which made the plant potentially unstable and easily susceptible to loss of control. These factors combined to produce a sudden power surge which resulted in violent explosions and the almost total destruction of the reactor. The graphite moderator in the reactor core then caught fire and contributed to a prolonged release of radioactive material.

It was quite a large release of radioactive material. Relative to a Nagasaki-sized explosion of 21 kilotons as a ground-burst, the Chernobyl accident released 890 times as much Cs^{137}. As such, it was equivalent to an 18.7 megaton ground-burst in terms of Cs^{137}, which is now the most significant radionuclide remaining from the accident. When fissile isotopes such as U^{235}, Pu^{239} and U^{233} split, two daughter atoms will be formed. Most of the daughter products tend to be about half the atomic number of the parent atom. If one daughter atom is much heavier than half the atomic number of the parent atom, then the other daughter atom will be much lighter. Thus the main radioactive nuclides, iodine and caesium, have an atomic number about half that of uranium. The isotopes of these elements, I^{131} and Cs^{137}, are responsible for most of the radiation exposure received by the general population. I^{131} has a half-life of eight days, decaying to xenon. Because of its short half-life, I^{131} is considered not to be a contaminant after two weeks and prophylactic dosing with potassium iodide can be discontinued two weeks after a nuclear incident. Nevertheless, since the accident there has been a real and significant increase of carcinomas of the thyroid among the population of infants and children exposed at the time of the accident in the contaminated regions of the former Soviet Union. The histology of the cancers has shown that nearly all were papillary carcinomas and they were particularly aggressive, often with prominent local invasion and distant metastases, usually to the lungs. This has made the treatment of these children less successful than expected. They were more prevalent in children aged zero to five years at the time of the accident and had a shorter latent period than expected. On the other hand the significant increase in cases of leukaemia, which had been so greatly feared, has not materialised. No increase of congenital abnormalities, adverse pregnancy outcomes or any other radiation induced disease in the general population, either in the contaminated regions or in Western Europe, resulted from the Chernobyl accident.

Contamination of agricultural land by Cs^{137} has been ameliorated by deep ploughing and the application of lime and potassium fertilisers. Deep ploughing by itself reduced plant uptake of Cs^{137} by two thirds. A successful method for reducing the contamination of livestock has been to add Prussian blue compounds to their diet. Prussian blue is the nontoxic ferrocyanide dye discovered in 1704. It binds the caesium in the gut and carries it off in the dung. The last restrictions of the sale of sheep from contaminated farms in Wales, 2,300 km west of Chernobyl, were lifted in mid-2012, 26 years after the accident.

In light of the above, radiological contamination from nuclear weapons and reactor excursions can be coped with and ameliorated. The main thing is to avoid exposure in the first couple of weeks as the short-lived isotopes decay. Beyond sensible precautions such as sheltering from fallout while it is happening, prophylactic dosing with potassium iodide would be necessary. A 14 day course of tablets is about eight dollars per person. If you want to survive with friends, the chemical supplier Nasco, based in Wisconsin in the United States, will sell you a half kilo bottle of potassium iodide crystals for US$96.75, which is enough to protect 360 people at $0.28 per head. That said, if there were to be a large scale nuclear exchange then a significant proportion of the country's farmland would become contaminated. There might be too much land contaminated for all of it to be taken out of production while waiting for the Cs^{137} to decay.

In that case, contaminated grain may have to be grown which will then be fed to animals with a Prussian blue supplement in their diet. Then, in turn, humans eating that radiologically contaminated meat will also have a Prussian blue supplement. Conversion of radiologically contaminated grain into animal protein, with Prussian Blue supplements in stages on the way, may be the only way to consume it.

What Australia should do

China is very likely to lose the conventional war it will start in the East and South China Seas. The communist leadership of China will not take this well. Losing the war would cause an enormous loss of prestige for the Chinese Communist Party and might cost the leadership their lives. When it becomes apparent that they are losing, China is likely to threaten Japan with a nuclear attack unless Japan stops fighting and hands over the entire Ryukyu island chain. Japan's only defence in that situation is the threat of retaliation with nuclear weapons by the United States. If that threat wasn't credible, Japan could very well be on the receiving end of some of China's three megaton warheads. China couldn't care less how many Japanese they killed.

President Obama came into office in 2008 promising to end nuclear weapons around the planet, including those of the United States. By 2015 that position was reversed and he has greenlighted the Iranian nuclear weapons program. The Iranian deal has provided the biggest push for nuclear weapons proliferation for decades. The existential question for the Japanese is, if China threatens Japan with nuclear annihilation, will President Obama threaten China with nuclear weapons in return? And actually use them against China if China wipes out Japan? President Obama's track record in office suggests that he wouldn't. Because he has not followed through on previous threats to use force, threats from this point wouldn't be credible. The nuclear umbrella under which Japan and Australia have sheltered is no longer there.

So the continued existence of the nation of Japan depends upon the mood swings and personality disorders of an American president.

And if not this president, perhaps some future one. That is not an acceptable situation for the Japanese people. In 2014, China became agitated by Japan's continued possession of 313 kilograms of weapons-grade plutonium that the United States had given Japan back in the 1960s to conduct nuclear research. That is enough to make about 50 nuclear bombs with 50 kiloton yields if given tritium boosting. Possession of that plutonium is the best guarantee of Japan's continued existence as a nation.

In an interview in January 2014, Colonel Liu Mingfu of China's National Defence University threatened Japan with nuclear attack. The significance of Colonel Liu is that he is the author of a tract called *The China Dream* which calls upon China to build the world's strongest military and move swiftly to topple the United States as the global "champion". In that same vein, the book says "China's big goal in the 21st century is to become world number one, the top power" and "If China in the 21st century cannot become world number one, cannot become the top power, then inevitably it will become a straggler that is cast aside". President Xi took the concept of *The China Dream* as his aspirational motto when he became leader in direct reference to the book which had been banned prior to his ascension. The further problem with the leaders of Chinese military thought is that they think in terribly simplistic terms For example, Colonel Liu is quoted as saying "America is the global tiger and Japan is Asia's wolf and both are now madly biting China. Of all the animals, Chinese people hate the wolf the most." And that "China was a peaceful nation but it would fight to the death if seriously attacked." This last statement points to China being at stage two of its threat signalling sequence as at early 2014.

With such a deranged neighbour, Japan should develop its own nuclear weapons and delivery systems and Australia should join it. The sooner the preparations begin for that, the better. The world is coming up to a big bifurcation point in history. China's base building in the South China Sea is only justified (to itself) if China goes on to declare an Air Defence Identification Zone over the whole South China Sea. If the United States does not nullify that, then the United States has given up on free passage on the high seas and has ceded Asia to China. The alternative is war which President Obama wouldn't have the stomach for. That would be the point for Japan to declare that it has developed nuclear weapons. Because, otherwise, China will be emboldened and go on to attack Japan in the Senkaku and Ryukyu Islands and threaten Japan with nuclear annihilation if it doesn't give up. Economic growth in Asia will then be over because of the diversion of GDP to military spending.

7

THE BROADER STRATEGIC CONTEXT

After China is defeated, the world won't return to normal. We have viewed the prosperity of the last fifty years as normal. In fact it was an abnormally benign period in human history due to a confluence of factors. The superpower nuclear standoff gave us fifty years of relative peace, we had cheap energy from an inherent over-supply of oil, grain supply increased faster than population growth, and the climate warmed because of the highest level of solar activity for eight thousand years.

All those trends are now reversing. World population was 2 billion in 1930. Now it is 7 billion, up 250 percent. World grain production was 481 million metric tons in 1930. Now it is 2.4 billion metric tons, up 392 per cent thanks to the green revolution pioneered by Norman Borlaug and others. Grain prices fell all through that period – up until the last few years. Developing country wheat yields peaked at 2.7 metric tons per hectare in 1996 and have plateaued thereafter. Developed country grain yields have plateaued from 2000. In the last decade the supply overhang has been absorbed, and now grain prices are running up. Meanwhile, each day sees another two hundred thousand people added to the world's population. As adults, each day's cohort will need sixty-six thousand metric tons of wheat per annum to keep body and soul together. That means that an additional 25 million metric tons of wheat production will be required to feed the world's population each and every year. Most of the world's population already spends a quarter to a half of their income on food. Thus rising food prices will have a severe impact on their discretionary spending, shrinking the market for goods and services.

If the climate was actually warming, vast areas of Canada and Russia could be put under the plough and contribute to the world's grain supply. But we know that the temperature of the planet has not risen for the last eighteen years (Climate is one subject that is studied intensely these days.) We can be almost as certain that a severe solar-driven cooling event is in train. Instead of the Northern Hemisphere grain belts moving north, they will be moving south. The U.S. Corn Belt will move towards the Sun Belt, just as the northern limit of American Indian corn-growing moved three hundred kilometres south between the Medieval Warm Period and the Little Ice Age. Grain production in Canada will become difficult. Norway's wheat production is already down 48 per cent from its peak in 2007 because of cold, wet summers. Total world grain stocks were about 330 million metric tons at year-end 2013, only 14 per cent of annual demand. As the cooling continues and worsens, nations dependent on imported grain are facing mass starvation.

As for world peace, the artificial nations created by the British and the French in the Middle East after World War I will devolve to their tribal components. That part of the world at least may go back to the Stone Age condition of 30 per cent of adult males dying violent deaths. Very few Middle Eastern countries produce all of their food requirements. Who will pay to keep them fed when grain becomes scarce and expensive? Added into that mix are the nuclear weapons of Pakistan (a future failed state) and the ones that Iran is intent on making.

The UN-EU establishment that gave us the global warming scare in order to establish a new world order (after the failure of communism) is well aware of the problem of food supply. The increasingly untenable global warming dogma is scheduled to be replaced by propaganda about "sustainability"; in fact, the switch is already underway. The campaign to control the world's food supply had its first official airing at a meeting of G20 agriculture ministers in Paris in 2011. It is telling that one of the bodies that suddenly decided to mount a campaign against "food waste" in late 2012 was the Environmental Protection Authority of New South Wales. The staff of this authority are people who are counting on spending the rest of their lives as social parasites in the field of environmentalism. They are evidently taking instruction to move on to this new field of agitprop from higher authorities in the UN-EU establishment, in a chain of influence (if not of command) outside their own government.

Global warming itself, as many others have noted, is the greatest swindle perpetrated in history. There was a pleasant warming that started in the mid-nineteenth century, but that warming is easily and completely explained by the highest rate of solar activity for eight thousand years. If you want to split hairs, the higher atmospheric carbon dioxide level of the last hundred years most likely contributed one tenth of one degree to that warming. The effect is lost in the noise of the climate system. But the invention of the global warming scare was necessary for the UN-EU establishment to negate the triumph of liberal democracy that Francis Fukuyama predicted in 1992. As a belief system, global warming gave its adherents some of the basics in spiritual nourishment – original sin, the fall from grace, absolution, redemption, and sacrifice, to name a few. It is hard to see the notion of sustainability providing the same quality of experience.

To optimise our strategy as a country, first of all we have to determine who we are, who are like us and with whom we have common interests, and who to avoid. The context of that used to be purely cultural, now economics is a good pointer. In the 1950s and 1960s it was widely assumed that the rising tide of modernity would "lift all boats". In fact the very opposite occurred. Only a few countries enjoyed sustained economic growth while the rest remained laggards. The former are basically the members of the Organisation for Economic Co-operation and Development (OECD) which was created on December 14, 1960. The OECD was the successor to the Organisation for European Economic Co-operation, which helped to administer the Marshall Plan through which the United States had helped to reconstruct war-battered Europe after 1948. The OECD is a forum of countries committed to democracy and the free-market economy. Significantly, Japan joined the OECD in 1964 and its nearest neighbour, South Korea, followed in 1996.

Economists expect that poor or "under-developed" or "developing" nations should grow faster than richer or "developed" economies, so that worldwide living standards can be expected to eventually converge. With a few exceptions, this has not happened. By the 1970s, it was evident that most non-OECD countries were still lagging the OECD in GDP per capita, and the gap actually widened as the decades passed. This gap caused existential angst in some Islamic countries. Moslems believed that their culture was as good as anybody's – in fact, the best – and yet they were lagging far behind. They rationalised this state of affairs with the notion that they were being oppressed – which led to terrorism against the West. Other nations, such as the sub-Saharan African states, simply took the aid money they were given while conditions continued to deteriorate from the levels set in the colonial period. For the first thirty-four years of the period of abundance, the world order was dominated by the battle between the OECD countries and the communist regimes. Eventually the stagnation of the communist regimes destroyed their moral authority and they collapsed. The sudden collapse of communism gave rise to the notion that the World had entered a golden age of harmony between countries and civilisations. That brief period of hubris ended on September 11, 2001 with Islamic suicide attacks on U.S. soil.

By then, what had been evident to angst-ridden Moslems for some time started attracting the attention of historians and strategists. It was evident that the OECD countries were doing a lot better than the rest of the planet and that that gap showed no signs of closing. So historians looked for reasons.

In 2001 Californian classicist and eminent historian Victor Davis Hanson published *Carnage and Culture*, an attempt to explain this state of affairs. Professor Hanson contended that there were six crucial elements in the Western's way of conducting wars that led to its arms prevailing for so many previous centuries:

1. Political freedom, which came from ancient Greece;
2. Civic militarism, which calls citizens to the defense of their property and society;
3. Decisive shock battle by disciplined infantry;
4. Technology and a scientific tradition;
5. Private property, which provides soldiers with a vested interest in the outcome;
6. Civilian audit and open dissent.[1]

Some of these features of Western civilisation – political freedom and private property, in particular – explain not only why the OECD countries are more effective militarily, but also why they are prosperous in other ways as well.

Professor Hanson's book was followed three years later by *The Pentagon's New Map*, written by former Pentagon strategist Thomas Barnett. Barnett noticed that the wars of the late twentieth century had occurred only in particular parts of the globe, while other parts were completely free of war. In explaining this state of affairs, Barnett made the following observations:

1. The world is divided into two groups: the Functioning Core with a high level of intra-group trade and the Non-Integrated Gap. The Core countries can be sub-divided into an Old Core comprised of North America, Western Europe, Japan, and Australia and a New Core of China, India, South Africa, Brazil, Argentina, Chile, and Russia. The Non-Integrated Gap consists of the Middle East, Southern Asia with the exception of India, most of Africa, and northwest South America.

2. Gap group countries can improve their economies and in turn reduce violence and terrorism by increasing their international trade. Failing that, the U.S. military is the only entity capable of maintaining order in the Gap countries and enforcing rules of conduct.

3. The U.S. military should take a holistic approach to war and consider war in the context of demographics, energy supply, trade, and other factors.

4. The U.S. military has two functions. As Leviathan, it uses overwhelming force to defend Core nations. As System Administrator, it concentrates on nation-building.[2]

Then in 2011 Harvard Professor Niall Ferguson published *Civilisation: The West and the Rest* in a bid to explain why Western countries have prevailed both militarily and economically. He attributes the divergence between the West and the Rest to what he called six "killer apps":

1. Competition. Europe was fragmented in the sixteenth century, and this created competition between countries, which in turn encouraged improvement.
2. Science. Most innovations in machinery and weaponry came from Europe.
3. Property rights. Professor Ferguson's view is that respect for private property rights encourages productivity and the accumulation of wealth.
4. Medicine. Western advances in vaccinations increased life expectancy.
5. Consumerism. Increased consumption grew trade and GDP.
6. The work ethic. Protestantism stressed hard work, saving, and reading.

Quite correctly, Professor Ferguson wrote that the greatest dangers facing us are probably not "the rise of China, Islam, or carbon dioxide emissions" but "our own loss of faith in the civilisation we inherited from our ancestors."[3]

There is one point of overlap between the assessments of Hanson and Ferguson – namely, the crucial importance of private property safe from seizure. The vital importance of private property to economic prosperity is best illustrated by the fate of countries that do not respect private property as well as they might. Argentina is the textbook example. At the beginning of the twentieth century, Argentina had a GDP per capita which was 80 per cent of the U.S. level. That relative measure promptly went into a long decline, which saw Argentinian GDP per capita fall to 30 per cent of the U.S. level by the end of the twentieth century. What happened? Argentina took the path of wealth redistribution and never recovered. Visitors to Buenos Aires report that it is a city of Europeans living in dirty poverty. To this day, the Argentinian Government is still seizing assets – simply because it can. Early in 2012 the Argentinian Government seized a 51 per cent stake in the Argentinian oil company YPF held by the Spanish oil company Repsol because YPF has what is perceived to be a valuable shale oil resource.

But there are places much more destitute than Argentina. Consider the riddle of why Haiti remains so desperately poor, as explained by Jeffrey Tucker in 2011:

> The answer has to do with the regime. It is a well-known fact that any accumulation of wealth in Haiti makes you a target, if not of the population in general (which has grown suspicious of wealth, and probably for good reason), then certainly of the government. The regime, no matter who is in charge, is like a voracious dog on the loose, seeking to devour any private wealth that happens to emerge. This creates something even worse than the Higgsian problem of 'regime uncertainty.' The regime is certain: it is certain to steal anything it can, whenever it can, always and forever.[4]

The great economic divide in the world is simply between those countries that respect private property and encourage individuals to accumulate wealth and those countries that make it difficult for individuals to accumulate wealth and property. The average GDP per capita of the former is four times that of the best GDPs of the latter. Respect for private property is the sole determinant of which group a country falls into. If private property is respected, all of the other economic virtues, such as a strong work ethic, come with it. Note that respect for private property, and thus membership of the OECD group, is not an exclusively Western virtue. Japan, South Korea, Taiwan, and Singapore all have high standards of living while neighbouring countries are far poorer. Any nation could choose to have a standard of living as high as that of the OECD average. It would just have to change the way it is run. Otherwise, there is an immutable barrier that stops per capita GDP rising beyond $10,000 per annum.

The division between Barnett's Core and the Non-Core (Non-Integrated Gap, in his terminology) is also respect for private property. The Non-Core countries are run as kleptocracies, either at the state level as in Argentina or at the warlord level in sub-Saharan Africa. But quite a few of the countries that Barnett counted in his Core are kleptocracies, as well, including Russia and China. The lack of respect for private property in those countries is a hard limit on their economic potential. One of

Barnett's useful observations is that Core Countries should support other Core countries and police the Non-Core countries. Foreign aid to Non-Core nations is completely wasted, as there can be no improvement in their standard of living without a change in their cultures—that is, in their attitudes toward private property. All foreign aid to such countries are Band-aid measures that simply allow the elites of those countries to steal more.

Threats to Australia won't stop after China is defeated. The first threat to deal with is internal. The prosperity of the last fifty years has led to a cult of self-loathing. It is not just a desire to make us poorer by using global warming as an excuse for international wealth redistribution, it is also a worship of things that are un-Australian and anti-Australian, such as illegal immigrants who don't share the values that got us into the OECD in the first place. As we have seen earlier in this chapter, there is a reason why poor countries remain poor and in fact are condemned to remain poor for all of eternity. Their culture won't allow work to be properly rewarded and wealth to be accumulated. Diluting Australian culture with these people simply makes us poorer and less secure. We don't have to have a debate about that though because Australia is full up. During droughts our grain production falls to the level of Australian demand. So, if our population continues to rise, during future droughts we would have to import food. We want to avoid that more than having to import fuel.

All our foreign aid is wasted, all $4 billion per annum of it. It is now 70 years after World War 2. If countries haven't got their act together by now, they will never get their act together.

The second thing we want to avoid is having anything to do with countries that are going to collapse due to starvation-driven collapse, which for starters is most of the Middle East and Africa. Consider Australia's involvement in the war in Afghanistan. Afghanistan is a country with violence at Palaeolithic levels, a misogynist religion and endemic corruption. It is surrounded by countries that are hostile to it and the West. In fact it seems to be made up of territory that is unwanted by the countries that surround it, as if the value from incorporating bits of Afghanistan would not be worth the pain of having the Afghani people who would come with it. It does not produce anything that the world wants in a positive way. Its major export is heroin, bringing misery and death to a large arc from Russia to Western Europe.

It sounds like a hopeless case of a country to do some nation-building in but that is not the worst of it. The modern history of Afghanistan is written in its wheat statistics. Back in 1960, there were nine million Afghans and they grew 2.3 million tonnes of wheat in that year. By the time the Russians invaded in 1979, wheat production had risen to 2.7 million tonnes with a further 200,000 tonnes being imported. There were then 13.7 million Afghans. Things did not go well in the latter half of the Russian period of occupation with wheat production falling to 1.8 million tonnes in 1989, the year they withdrew. Nevertheless, population growth rate did not fall below two per cent per annum while the Russians were in charge with the population growing to 16.9 million in the year they left.

Population growth increased to 2.6 per cent per annum under the Taliban. By the time the United States' turn at running the country began in 2001, the number of Afghanis in existence had increased to 22.8 million. Thirteen years further along, there are now 32 million Afghanis, an increase of just over nine million. How many Afghanis have died in the conflict since the United States-led coalition entered the country? It may be as high as fifteen thousand, with two-thirds of those having been killed by the Taliban. The ratio of creation of new Afghans by birth to deaths of Afghans in the ongoing conflict is over 600 to 1. What is the carrying capacity of the country? Under ideal conditions, aided by the warmest climate for 800 years, it is perhaps 13 million people. What keeps the excess above that figure alive is imported grain which for the last few years has settled down to a rate of about two million tonnes per annum.

Afghanistan's population growth rate is now 2.4 per cent per annum. At that rate it is doubling every 29 years. By 2030 there will be 46.6 million Afghans. To keep body and soul together, the increased population from the 2014 level will require a further 5 million tonnes per annum of imported wheat. Can anyone think of where the money might come from to pay for that wheat, if the wheat can be found at that time in the first place? Eventually the international aid donors will get sick of paying for the ungrateful and irredeemable wretches. Once wheat imports fall below what is required to keep the population quiescent, rioting and social breakdown will follow. The starving urban populations spread out into the countryside, devouring what they can, including the seed grain for the next crop. Population falls to a fraction of the country's carrying capacity after the death of some 30 million Afghanis. There is no force on Earth that can stop something like this from happening. There is no limitless supply of money and no limitless supply of grain that can overcome a population doubling period of 29 years.

The last time Syria attacked a civilised nation was the Yom Kippur War of 1973 in which it and Egypt launched a surprise attack on Israel. At the time, Syria had a population of seven million and Egypt 38 million. It was also about the last time that both countries could feed themselves from their own agricultural efforts. Their populations are now 22 million and 83 million respectively with all the increase in population from 1973 fed with imported grain. This is true of the whole Middle East – North Africa (MENA) region. This is shown in Figure 32 going from Morocco in the west to Afghanistan in the east.

But grain yields in most countries have plateaued since 2000 and grain prices have started rising again. That effect will be accelerated by the solar-driven global cooling that has started. At some stage the cost of keeping everyone fed will overwhelm one of the MENA countries and it will collapse in mass starvation. There will then be a mad scramble around the world to stockpile grain, sending prices yet higher. In turn that will set off a domino effect in the region. Using an animal model of population collapse (the snowshoe hare and lynx), populations might fall to 10 per cent of carrying capacity – back to levels last seen 200 years ago.

Israel imports most of its grain requirements as do all its neighbours. The big difference is that Israel has a GDP per capita of $30,000-odd which is at least ten times that of its neighbours. Israel could afford a much higher grain price. That country also is the most efficient desalinator of seawater on the planet. With a cost of $0.52 per cubic metre, it is able to grow commercial crops using desalinated seawater. For Israel to survive from here, all it has to do is out-wait its neighbours. In the good old days, a large population meant a country could have a large army. These days it means the ongoing drag of having to feed a lot of unproductive people with every missed grain shipment a potential disaster.

There are now four million Syrian refugees outside Syria and those people cost over $8 billion per annum to be kept alive. That works out at about $2,000 per capita or $5.50 per day. That is a useful metric to apply to other situations. For example, control of Yemen recently passed from groups friendly to Saudi Arabia to a group backed by Iran. Saudi Arabia used to pay for its southern neighbour to be kept fed. Now that has become Iran's responsibility.

There are 24.4 million Yemenis living on a patch of mostly desert that could support perhaps 10 per cent of that number from its own agriculture. At $5.50 per day, Yemen may take of the order of $100 million each and every day to be kept fed. That cost has been added to all the other fights that Iran has taken on.

There are some bizzare arrangements in the Middle East. For example, while Australian personnel and assets are being used against Islamic State, a UN agency is sending food into Islamic State-controlled areas. If we weren't feeding them, Islamic State would collapse quickly.

A lot of the folktales we were told as children involved starving peasant children begging for scraps of bread. That is what happened in Europe during the creation of our folk memory. People spent a lot of their lives on an involuntary calorie-restricted diet. Starvation is what stopped the human population blowing out to 10 billion thousands of years ago.

As a civilisation, we have largely forgotten about starvation because of two technical advances. The Haber Bosch process enables coal or natural gas to power the conversion of atmospheric nitrogen to fertiliser (the source of half the protein on our plates) and the green revolution pioneered by Norman Borlaug. Meanwhile, population growth in a lot of places is still galloping along. For example the population of the West African country of Gambia is doubling every 26 years. Gambia is importing food now so all the growth in population, while it continues, will have to be fed with imported grain. It is a game of musical chairs that will end in tears when the music stops. But nobody cares much what happens in West Africa. Perhaps the supply of cocoa beans for making chocolate will be interrupted. At the time of West Africa's inevitable starvation and population collapse, everyone else on the planet will be

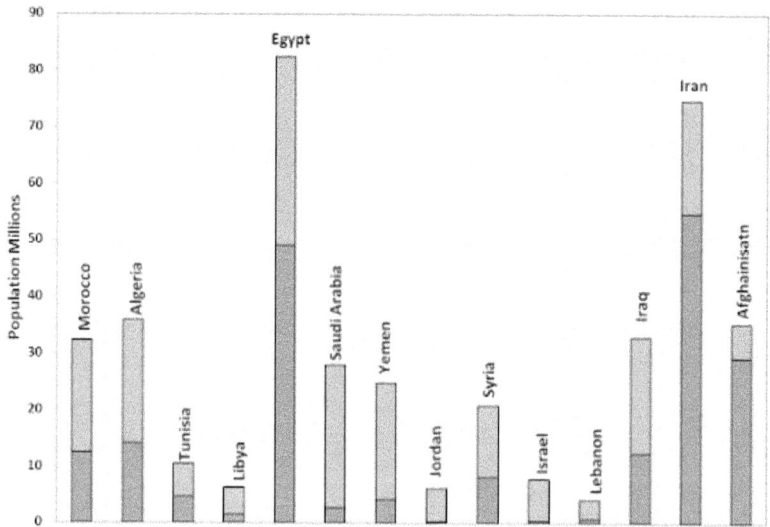

Figure 32: Domestic Grain Production and Grain Imports for the MENA Region

The size of each bar is the country's population in millions. The light part is the proportion fed from domestic production and the dark part is the proportion kept alive with grain imports. Arguably Norman Borlaug and his green revolution allowed this situation to come about. World grain production outran population growth up to about a decade ago and grain became the cheapest it has been in history. Feeding these growing populations has been very cheap for regimes that subsidise bread to keep their populations quiescent. At the current price of wheat of $330 per metric ton and per capita consumption of 300 kg per annum, it only costs $0.27 per day in grain to keep someone alive.

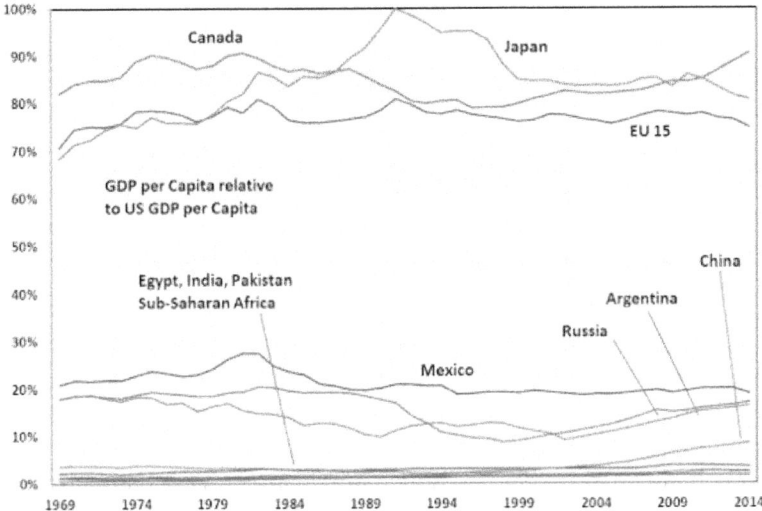

Figure 33: GDP per Capita relative to US GDP per Capita

The world is divided into two as shown by this graph of GDP per capita for countries and regions relative to US GDP per capita. The data comes from the US Department of Agriculture which tries to predict how much food countries will import. There are no countries that exist in the gap between the OECD and the rest. There has been no convergence of the non-OECD countries towards the standard of living of the OECD countries in the 25 years of data shown in this graph. Therefore the only way non-OECD countries can increase their standard of living is to improve their cultures.

more focussed on their own survival and getting bread into their own children's mouths.

We have digressed – back to the MENA region. The fate of the region is a collapse event with the deaths of several hundred million people when the grain ships stop arriving. Democracies or any other type of regime can't be installed with any hope that they will endure. The Syrian cities that are now rubble are just another layer (funded by fossil fuels) of civilisation on top of the ruins that preceded them. The whole region will end up like that. Don't get involved. Don't get fond of anyone in the region or choose one over the other. If you try to fix one of their problems you get to own all of their problems, and all of their problems are intractable. Don't even bother to learn their names. The reasons why some of them might be upset with the rest of the world count for nothing. Don't parlay with them because their undertakings mean nothing in the long term. And when they stop paying for their grain, don't send them grain. Don't let them into this country, and don't pretend that their civilisations, with all their repellent customs, are worth visiting. That is a list of seven don'ts. That is all that is needed to be known.

The Romans had similar problems in the region two thousand years ago. Their response to being attacked was to wipe out the entire population of the country or tribe that attacked them. We don't have to go that far these days. All we have to do is remove the errant leadership of the offending entity.

In the absence of a deterioration in climate in the interim, global population will reach the limit of potential global agricultural production by 2035. Solar activity is now declining from its recent 8,000 year peak of activity with the consequence that deterioration of agricultural growing conditions will follow. Cold-driven famines over the last few hundred years are well documented. Severe cold in the 1690s killed 30 per cent of the population of Finland, and lesser percentages of other countries from France to Sweden. In Ireland, 20 per cent of the population died in 1740, one hundred years before the more famous potato famine. In 1816, the Swiss were eating their cats, dogs, rats and horses due to the combination of a solar-driven cold period called the Dalton Minimum and the eruption of Mount Tambora in Indonesia the year before.

The United States is the world's largest agricultural exporter and keeps body and soul together for a lot of the world. There are several factors that affect agricultural productivity in the United States:

1. Productivity is directly proportional to temperature in the mid-latitudes. For the Corn Belt it is 10 per cent per 1.0°C of average annual temperature. Corn requires temperatures above 10°C, wheat above 4°C. Where it is too cold to get a corn crop off in a season, wheat is grown. Where too cold for wheat, rye and oats are possible.

2. As growing conditions move south, some formerly productive land will be abandoned.
3. Regions in the south in which the length of the growing season allows double cropping will be reduced to one crop per season.
4. Wheat production could go up by a switch from hard summer wheat to winter wheat which has a lower protein content and makes lower quality bread.
5. Production will respond to higher prices – several farmers from the Midwest have told me that they could have a big increase in production with the right price signal.
6. Large areas of land in the southeast and northeast US that are currently non-competitive with the Corn Belt could be brought into production with the right price signal.
7. Also with the right price signal, a lot of food could be grown residentially. During WW2, 40 per cent of US vegetable production was from domestic plots. The appropriate high protein plant crop is soybeans which are 12 per cent protein in the green bean. You can grow potatoes easily enough but they are only 2 per cent protein.

All things considered, the production decline for US agriculture could be 8 per cent per 1°C. A fall of 3°C and the United States would be out of export markets for agricultural products, with the same true of most mid-latitude grain exporters. This will have profound geopolitical implications – namely, starvation and collapse for countries that import food. That's for next decade. This decade, once the temperature decline is widely apparent, currently importing countries around the world will rush to stockpile, bringing forward the price effect of scarcity.

If this sounds like the reverse echo of the global warming crowd, how often have leftist wealth-redistributors been right about anything? Given their track record, the exact opposite is the more likely outcome. We should be ever thankful to them though. If it wasn't for their melodramatic predictions attracting honest scientists into the climate science field, humanity would be sleepwalking into the climatic and agricultural disruption that is coming. We will still have the consequent famine and death but we will know what's causing it at the time. We are on the edge of that abyss. A vale of tears awaits.

Australia, with its fragile agriculture due to rainfall variability, should avoid the possibility of starvation in this country by limiting immigration, legal and illegal. Prospective migrants should be told to adopt Australian culture in their own country as their best hope of achieving a higher standard of living.

8

FUNDING THE INCREASED DEFENCE EFFORT

It is an old saw that amateurs talk strategy while professionals talk logistics. Logistics encompasses the cost of the whole force structure as well as supplying it. As will be shown in this book, Australia needs to at least double its defence expenditure if it doesn't want to have terms dictated to it in the next major conflict. So where is the money for that to come from? Let's put that to bed at the outset. There is plenty of money for defence. It just happens to be misdirected at the moment to other departments.

One of the reasons the Coalition won the 2013 federal election was because they pointed out that Australia's spending on defence had fallen to 1.38 per cent of GDP, the lowest level since 1938. They promised to spend at least 2 per cent of GDP on defence. Upon becoming the government, they have found that hard to do. The current promise is to increase defence spending by 3 per cent per annum until the 2 per cent of GDP figure is achieved. Those good intentions have been overtaken by events. China's aggression dictates that we should at least double defence spending instead.

The previous government had promised many things too. The 2009 Defence white paper was entitled "Defending Australia in the Asia Pacific Century: Force 2030". This is the document that promised twelve new submarines amongst other good things for Australia's defence. It had a short shelf life because eight days later the defence budget was slashed. Labor kept cutting the Defence budget with a further cut of 5 per cent in 2010 and 10.5 per cent in 2011. Prime Minister Rudd had no intention of funding any submarines in any quantity but did not want to be accused of being soft on defence. His successors have continued to pay lip service to defence.

Wars are traumatic things. Defence expenditure peaked at 40 per cent of GDP in 1943 during World War II, during which a few hundred Australians were killed on Australian soil – from Sydney Harbour to Darwin and round to Port Hedland. So after that war, generally accepted wisdom was that defence expenditure should be a minimum of three per cent of GDP. Given the state of the world today with four civil wars in the Middle East-North Africa region and the showdown coming in East Asia, getting to a minimum of three per cent of GDP on defence would be a good idea - the sooner the better.

Where is the money to come from? All taxes suppress economic activity so we don't want to do it by increasing any form of tax or by introducing new ones. The biggest single lump of Federal Government expenditure is Social Security and Welfare at about $120 billion and 9 per cent of GDP. Defence is currently one sixth that.

We all know that we have had a boom since the turn of the millennium with Australia's GDP up 46 per cent since 2000. Population grew 21 per cent over the same interval so, on average, we are about 20 per cent better off than 14 years ago. It turns out that welfare recipients are also about 20 per cent better off than in 2000 with per capita welfare spending rising (in 2010 dollars) from $4,483 to $5,390 in 2013. That 20 per cent increase now accounts to about $20 billion per annum. A good part of the increase was during the Howard years – in fact the percentage growth in per capita welfare spending dropped in the Rudd/Gillard years. Howard just wanted to be re-elected and was prepared to spend the country's money to do so.

So there's plenty of money for defence – it's in the welfare budget. The defence budget could be near doubled simply by taking per capita welfare spending back to the level of 2000 and handing the savings over.

Would anybody suffer if the welfare budget was cut back? Let's take the case of the Cairns mother who allegedly stabbed her seven children to death in late 2014. Those children were by five different fathers. That's not the problem, the problem is that there is no doubt that taxpayers were paying to keep body and soul together for those seven children. If the mother had to pay for the children herself, she might not have had any at all. She might have had a job and contributed to society instead of just being a social parasite.

Scott Morrison started his post as Minister for Social Security by promising to reform the heavily rorted disability pension system. There are some 800,000 on the disability pension. A high proportion of these are able to drive a car, bash police and each, go fishing and so on. We often hear about disability pensioners when their active lifestyle resulted in a contretemps with another party, most recently the recipients of disability payments who have overcome their infirmities to take up arms against civilisation in Syria and Iraq.

In fact Australia now spends more upon Aboriginal welfare[1] than it does on Defence. In 2014, our spending in that sector was $30.3 billion. Most of that money would have been spent on white paper-shufflers, particularly white lawyers. Spending on Aboriginal welfare could probably be cut to $5 billion per annum without affecting the quality of life of anybody who was more than half aboriginal. As Albert Einstein said, insanity is doing the same thing over and over again and expecting different results. The country's spending on Aboriginal welfare is insane. If it is never going to have an effect, we might as well stop doing it and reallocate the funds to something that is vitally important: our defence.

Between the growth of welfare spending post-2000 to fund John Howard's re-election and the misapplied funds to Aboriginal welfare, an additional $50 billion a year is available that should be spent on defence. And we will need it all.

Spending that money won't cure all our ills though. For it to be applied effectively, there has to be some attitude changes at several levels. The Defence bureaucracy is just as inefficient and myopic as any of its Canberra colleagues, though the reforms being instituted as a result of the Peever Report[2] might effect an improvement. There is also a malaise at the national level. One symptom of that is the continuing preoccupation with global warming and the willingness to squander billions of dollars on fighting that phantom menace. The science of global warming has been discredited and the climate itself has refused to play along by temperatures remaining flat for almost two decades. In fact we can now say that no child alive has experienced global warming. And they are unlikely ever to in the rest of their lives given well-regarded predictions for solar activity. The point of this is that if we are bamboozled by something as simple as climate science in which we have had decades to sort right from wrong, what hope do we have of getting right more complex problems relating to national security?

But the prime example of the disconnect between reality and the way the country is run is the reaction to some foreign volcanoes. In June 2011, the Puyehue volcano in southern Chile erupted and ejected some volcanic ash into the atmosphere. It wasn't a big eruption. The coarser ash falls close to a volcano and some of the very fine material gets to the upper atmosphere and is carried by high level winds, being diluted with distance. Aircraft have only had trouble with volcanic ash when they have flown through the plume above the volcano. Nevertheless, some of the very fine ash got carried by stratospheric winds 20,000 kilometres to above Australia. Its presence was detected by satellites. Qantas grounded some scheduled aircraft flights because of the presence of volcanic ash in barely detectable amounts on satellite imagery.

Previously Qantas had not worried about volcanoes because the satellite imagery to see the ash plume had not been available. Qantas was aping the Europeans the year before who had shut down European air travel because of the ash plume from the Eyjafjallajokull eruption in Iceland. At least the Europeans subsequently instituted a safe limit of 4 milligrams of ash per cubic metre of air to avoid needless shutdowns of air space. There is no science involved when Australian air space is shutdown.

More recently, the Sangeang Api volcano in Indonesia erupted in June 2014 and the stupidity was repeated. The federal Department of Transport shut down Darwin airport, 1,400 kilometres from the volcano. The Minister for Transport mentioned the possibility of shutting down Brisbane airport, 4,000 kilometres from the volcano. Meanwhile, what do the Indonesians, the Filipinos and the Japanese do about their erupting volcanoes? They simply fly around them and no harm is done. You might think that the Transport Minister and others in the system might be curious about the Indonesian response to erupting volcanoes. But no – no curiosity at all. The Indonesians must rightly think that we are idiots in a hysterical schoolgirl sort of way.

The Australian response to erupting volcanoes so far away is indicates enormous stupidity and a lack of any sense of how the physical world works. It is like we have gone back to believing in magic and fairies at the bottom of the garden. We may keep re-electing really stupid people to govern us but the way history works is that stupid people are swept away by events and harsh realities take their toll on those who voted for them. In the meantime be very afraid because the people who believe in global warming and the magical properties of volcanic ash over enormous distances are still in charge of national security.

This book is an alarm about Chinese aggression. It details what we have to protect us, how those assets will be used, the deficiencies of some of our equipment and the weapons we should be buying instead. It is also a plea to get involved. Plato said that the penalty good men pay for indifference to public affairs is to be ruled by evil men. We are ruled by people who are not so much evil as indifferent themselves. To survive, let alone prosper, we are going to have to overcome our collective indifference. To improve our collective outcome, become part of that process.

Postscript

First of all, let's put the shopping list together. Following is not an exhaustive list of things that Australia should be doing to get its defence house in order:

1. Park up our 59 turbine-powered Abrams tanks and get 800 of the Leopard 2 tank weighing 68 tonnes, or the South Korean K2 Black Panther tank weighing 55 tonnes, or the Japanese Type 10 tank weighing 44 tonnes.
2. For the infantry fighting vehicle, get 800 of the CV90 weighing 28 tonnes.
3. Develop a nuclear weapons capability jointly with Japan.
4. Acquire 600 of the South Korean K9 Thunder 155 mm self-propelled howitzer.
5. Buy a further 250 M117 155 mm towed howitzers.
6. Acquire 90 of the HIMARS GPS-guided rocket system with reloads.
7. Sell our 47 MRH 90 helicopters and replace them with 60 UH-60M Black Hawk helicopters.
8. Acquire 100 M-28 Skytruck transport aircraft.
9. Develop eight kilowatt battlefield lasers mounted in Bushmaster vehicles.
10. Acquire 12 to 24 Soryu class submarines from Japan, which may include vessels built in Adelaide.
11. If possible, get an allocation from the *Virginia* class submarine production lines in the United States of at least four.
12. Failing the *Virginia* class, get an allocation from the *Barracuda* class production line in France and put up with the 10 year refuelling interval instead of the 33 life without refuelling of the *Virginia* class.

13. Embark on continuous build in Adelaide of 3,500 tonne version of the *Formidable* class frigate used by the Singapore Navy. It doesn't matter if at some stage next decade there is an excess of such boats. In a war, for every three of our ships sunk we should be able to put together at least one crew to man ships out of reserve.
14. Provide covered berthing for the submarines at HMAS *Stirling* in Perth.
15. Build a naval base in Exmouth Gulf with covered berthing for submarines.
16. Build a naval base on Cape Peron in Shark Bay with covered basing for submarines.
17. Start replacing the *Armidale* class patrol boat with a military version of the *Cape* class patrol boat.
18. Build a submarine base with covered berthing in Double Bay south of Bowen in Northern Queensland.
19. Have a submarine tender based in Double Bay ready to be forward deployed to Manus Island in New Guinea.
20. Limit the purchase of F-35s to the two aircraft contracted for.
21. Acquire 222 Gripen E fighter aircraft from Saab, with the first of these built on the production line in Sweden and the rest in Australia.
22. Develop Australian cruise missile production capability.
23. Acquire 30 used Boeing 737 aircraft and convert them to bombers carrying anti-ship cruise missiles.
24. Buy 30 C-295 Persuader aircraft.
25. Buy four US-2 seaplanes from ShinMaywa of Japan.
26. Build JORN radars on the Broome Peninsula north of Broome and Bathurst Island north of Darwin.
27. Double the size of our aircrew training fleet.

28. Start building coal-to-liquids plants to make Australia self-sufficient in diesel, jet fuel and petrol.
29. Build and fill 100 million barrels of jet fuel and diesel storage.
30. Seal dirt airstrips across northern Australia to provide austere, dispersed basing for the Gripen aircraft.
31. Replace the diesel engines in the *Collins* class submarines during their next full cycle docking.

Above are the physical assets. On the cultural side, stop the degradation of the services in the form of selling off bases such as the Leeuwin Barracks in Perth and attempting to shoehorn displaced units into other bases. At another level it is all about restoring silverware to the officers' messes around the country. If we want an exceptional level of dedication from our armed services, we need to treat them with respect. Things like the silverware in the officers' messes are a cost-effective way of doing that.

Appendices

Appendix 1: The ANZUS Treaty

SECURITY TREATY BETWEEN AUSTRALIA, NEW ZEALAND, AND THE UNITED STATES OF AMERICA

THE PARTIES TO THIS TREATY,

REAFFIRMING their faith in the purposes and principles of the Charter of the United Nations and their desire to live in peace with all peoples and all Governments, and desiring to strengthen the fabric of peace in the Pacific Area,

NOTING that the United States already has arrangements pursuant to which its armed forces are stationed in the Philippines, and has armed forces and administrative responsibilities in the Ryukyus, and upon the coming into force of the Japanese Peace Treaty may also station armed forces in and about Japan to assist in the preservation of peace and security in the Japan Area,

RECOGNIZING that Australia and New Zealand as members of the British Commonwealth of Nations have military obligations outside as well as within the Pacific Area,

DESIRING to declare publicly and formally their sense of unity, so that no potential aggressor could be under the illusion that any of them stand alone in the Pacific Area, and

DESIRING further to coordinate their efforts for collective defence for the preservation of peace and security pending the development of a more comprehensive system of regional security in the Pacific Area,

THEREFORE DECLARE AND AGREE as follows:

Article I

The Parties undertake, as set forth in the Charter of the United

Nations, to settle any international disputes in which they may be involved by peaceful means in such a manner that international peace and security and justice are not endangered and to refrain in their international relations from the threat or use of force in any manner inconsistent with the purposes of the United Nations.

Article II

In order more effectively to achieve the objective of this Treaty the Parties separately and jointly by means of continuous and effective self-help and mutual aid will maintain and develop their individual and collective capacity to resist armed attack.

Article III

The Parties will consult together whenever in the opinion of any of them the territorial integrity, political independence or security of any of the Parties is threatened in the Pacific.

Article IV

Each Party recognizes that an armed attack in the Pacific Area on any of the Parties would be dangerous to its own peace and safety and declares that it would act to meet the common danger in accordance with its constitutional processes.

Any such armed attack and all measures taken as a result thereof shall be immediately reported to the Security Council of the United Nations. Such measures shall be terminated when the Security Council has taken the measures necessary to restore and maintain international peace and security.

Article V

For the purpose of Article IV, an armed attack on any of the Parties is deemed to include an armed attack on the metropolitan territory of any of the Parties, or on the island territories under its jurisdiction in the Pacific or on its armed forces, public vessels or aircraft in the Pacific.

Article VI

This Treaty does not affect and shall not be interpreted as affecting in any way the rights and obligations of the Parties under the Charter of the United Nations or the responsibility of the United Nations for the maintenance of international peace and security.

Article VII

The Parties hereby establish a Council, consisting of their Foreign Ministers or their Deputies, to consider matters concerning the implementation of this Treaty. The Council should be so organized as to be able to meet at any time.

Article VIII

Pending the development of a more comprehensive system of regional security in the Pacific Area and the development by the United Nations of more effective means to maintain international peace and security, the Council, established by Article VII, is authorized to maintain a consultative relationship with States, Regional Organizations, Associations of States or other authorities in the Pacific Area in a position to further the purposes of this Treaty and to contribute to the security of that Area.

Article IX

This Treaty shall be ratified by the Parties in accordance with their respective constitutional processes. The instruments of ratification shall be deposited as soon as possible with the Government of Australia, which will notify each of the other signatories of such deposit. The Treaty shall enter into force as soon as the ratifications of the signatories have been deposited.

Article X

This Treaty shall remain in force indefinitely. Any Party may cease to be a member of the Council established by Article VII one year after notice has been given to the Government of Australia, which will inform the Governments of the other Parties of the deposit of such notice.

Article XI

This Treaty in the English language shall be deposited in the archives of the Government of Australia. Duly certified copies thereof will be transmitted by that Government to the Governments of each of the other signatories.

IN WITNESS WHEREOF the undersigned Plenipotentiaries have signed this Treaty.

DONE at the city of San Francisco this first day of September, 1951.

For Australia: Percy C Spender

For New Zealand: C A Berendsen

For the United States of America: Dean Acheson, John Foster Dulles, Alexander Wiley, John J Sparkman

Appendix 2: Japan-Australia Joint Declaration on Security Cooperation

The Prime Ministers of Japan and Australia,

Affirming that the strategic partnership between Japan and Australia is based on democratic values, a commitment to human rights, freedom and the rule of law, as well as shared security interests, mutual respect, trust and deep friendship;

Committing to the continuing development of their strategic partnership to reflect shared values and interests;

Recalling their on-going beneficial cooperation on regional and global security challenges, including terrorism and the proliferation of weapons of mass destruction and their means of delivery, and human security concerns such as disaster relief and pandemics, as well as their contributions to regional peace and stability;

Recognising that the future security and prosperity of both Japan and Australia is linked to the secure future of the Asia-Pacific region and beyond;

Affirming their common purpose in working together, and with other countries through such fora as Asia Pacific Economic Cooperation (APEC), the ASEAN Regional Forum (ARF), and the East Asia Summit (EAS), to achieve the objective of a prosperous, open and secure Asia-Pacific region, and recognising that strengthened bilateral security cooperation will make a significant contribution in this context;

Committing to increasing practical cooperation between the defence forces and other security related agencies of Japan and Australia, including through strengthening the regular and constructive exchange of views and assessments of security developments in areas of common interest;

Committing to working together, and with others, to respond to new security challenges and threats, as they arise;

Affirming the common strategic interests and security benefits embodied in their respective alliance relationships with the United States, and committing to strengthening trilateral cooperation, including through practical collaboration among the foreign affairs, defence and other related agencies of all three countries, as well as through the Trilateral Strategic Dialogue and recognising that strengthened bilateral cooperation will be conducive to the enhancement of trilateral cooperation;

Desiring to create a comprehensive framework for the enhancement of security cooperation between Japan and Australia;

Have decided as follows:

Strengthening Cooperation

Japan and Australia will strengthen their cooperation and consultation on issues of common strategic interest in the Asia-Pacific region and beyond. This includes cooperation for a peaceful resolution of issues related to North Korea, including its nuclear development, ballistic missile activities, and humanitarian issues including the abduction issue. Japan and Australia also recognise the threat to peace and stability in the Asia-Pacific region and beyond posed by terrorism and will further strengthen cooperation to address this threat.

Japan and Australia will also strengthen their cooperation through the United Nations and other international and regional organisations and fora through activities such as peacekeeping and humanitarian relief operations. Japan and Australia will work towards the reform of the United Nations, including the realization of Japan's permanent membership of the Security Council.

The cooperation will be conducted in accordance with laws and regulations of each country.

Japan and Australia will deepen and expand their bilateral cooperation in the areas of security and defence cooperation with a view to enhancing the effectiveness of their combined contribution to regional and international peace and security, as well as human security.

Areas of Cooperation

The scope of security cooperation between Japan and Australia will include, but not be limited to the following:

- (i) law enforcement on combating transnational crime, including trafficking in illegal narcotics and precursors, people smuggling and trafficking, counterfeiting currency and arms smuggling;
- (ii) border security;
- (iii) counter-terrorism;
- (iv) disarmament and counter-proliferation of weapons of mass destruction and their means of delivery;
- (v) peace operations;
- (vi) exchange of strategic assessments and related information;
- (vii) maritime and aviation security;
- (viii) humanitarian relief operations, including disaster relief;
- (ix) contingency planning, including for pandemics.

As part of the above-mentioned cooperation, Japan and Australia will, as appropriate, strengthen practical cooperation between their respective defence forces and other security related agencies, including through:

- (i) exchange of personnel;
- (ii) joint exercises and training to further increase effectiveness of cooperation, including in the area of humanitarian relief operations;
- (iii) coordinated activities including those in the areas of law enforcement, peace operations, and regional capacity building.

Implementation

Japan and Australia will develop an action plan with specific measures to advance security cooperation in the above areas.

Japan and Australia will further strengthen the strategic dialogue between their Foreign Ministers, on an annual basis.

Japan and Australia will build on their dialogue between Defence Ministers, on an annual basis.

Japan and Australia will enhance joint Foreign and Defence Ministry dialogue, including through the establishment of a regular Ministerial dialogue.

Signed at Tokyo this 13th day of March, 2007

Shinzo Abe	John Howard
Prime Minister of Japan	Prime Minister of Australia

Appendix 3: The case for a new Australian grand strategy

Ross Babbage

It's time for Australians to come to grips with their more troubling security outlook and debate how best to strengthen their deterrence and defensive capabilities. That's the core message in my essay *Game Plan: The Case for a New Australian Grand Strategy*, which was recently published by the Menzies Research Centre. Let me summarise a few of the key points for fellow readers of *The Strategist*.

What are some of the major changes in Australia's security outlook? Let me touch on just six.

First, Australians need to appreciate that the Western allies no longer dominate the Western Pacific economy. The global balance of power is shifting markedly.

Second, we need to appreciate the unusual nature and massive scale of China's military expansion. In particular, the People's Liberation Army is not trying to match the US ship-for-ship, aircraft-for-aircraft or tank-for-tank. China's asymmetric strategy rather places emphasis on building surveillance, missile, submarine, counter-satellite, cyber, and other capabilities that place the Western allies' concentrated bases at risk, conventional force structures, and vulnerable logistic systems.

Beijing's assertive international strategy is an even more fundamental change in Australia's security environment. From a largely introverted posture 20 years ago, the regime in Beijing is now championing nationalist causes offshore, largely to reinforce its domestic legitimacy. That has driven the Chinese to launch aggressive cyber operations against Australia and its close allies, engage in dangerous confrontations with Japanese forces in the East China Sea, harass American ships and aircraft and dredge up new islands to reinforce its assertion of sovereignty over 80 per cent of the South China Sea.

A fourth major change is that the US has morphed into a less confident and more hesitant ally. While Washington has announced a rebalancing of its forces to the Pacific, with some 60 per cent of US naval and air assets planned for deployment to the theatre by 2020, the resources to implement the rebalance have been limited. US readiness levels have fallen and Washington has shown itself to be easily distracted by crises elsewhere.

In short, the strategic tides in the Indo–Pacific have been flowing against the US and its allies. Symptomatic of the shift is that during the last decade China's defence spending quadrupled whereas US defence spending rose by a total of only 12%.

For Australia, the strategic implications of these developments are profound. During the Cold War, the centre of superpower competition, tension, and potential conflict was in Central Europe. Australians became used to being located in a strategic backwater. But now the situation is markedly different. Whether we like it or not, we now find ourselves close to the centre-stage of major power competition, international tensions and potential conflict.

Japan, India, and most countries in Southeast Asia broadly share Australia's unease and in differing ways are taking steps to reinforce their security.

What should be the core themes of Australia's grand strategy in this new era?

There's clearly a need to accelerate the strengthening of Australia's independent defence capabilities. At a minimum, this will mean boosting defence spending to two per cent of GDP by 2022, as currently planned.

Given the heightened security concerns of most of our neighbours, there's a strong case for further strengthening Australia's security partnerships in the Indo–Pacific to reinforce regional resilience and confidence.

Several specific initiatives deserve consideration that have the potential to turbo-charge the Australia–US alliance.

First, Australia could reinforce American, Singaporean, and other regional efforts to enhance the maritime domain awareness of partner countries in the Indo-Pacific. Working with regional partners to develop a common operational picture of the maritime domain would strengthen local defences, reinforce practical cooperation and help to bolster regional confidence.

Second, the United States and most of Australia's other security partners currently have difficulties accessing a comprehensive network of military exercise and range facilities in the Indo–Pacific. This is a growing problem for the United States as it looks to position the bulk of its naval and air forces to the theatre. Relocating extra forces forward is one thing, but maintaining them in this theatre in a high state of readiness is another challenge altogether.

Australia already possesses exercise and range facilities that are large, relatively uncluttered, and feature diverse air, sea, and land environments. Australia could readily establish an Indo–Pacific Exercise and Range Complex that could be made available to its close allies and security partners on agreed terms and conditions. This would make extended American force deployments to Australia and its surrounding region much easier, more effective, and less expensive.

Third, the major changes under way in the Indo–Pacific are already forcing Australian and US defence planners to re-think many assumptions about future operations in this theatre. In consequence, there would be benefit in forming a small, high-quality Australia–US Strategic Planning Group.

Fourth, there is scope for Australia to exploit its strong track record for quality intelligence products, its geo-strategic location, its high quality workforce, and its technological sophistication by investing strongly to become the intelligence hub for close allies in the Indo–Pacific.

Fifth, Australia already hosts a number of facilities that support US and allied space programs. There's scope for Australia to do more in this field to strengthen allied command resilience and operational capabilities in the Indo–Pacific.

Finally, the Pentagon is facing dilemmas as it contemplates basing modes in the Western Pacific for the coming half century. US basing in the region is currently over-concentrated and operationally constraining.

Canberra could assist greatly. Lowy Institute polling reveals that more than 80 per cent of Australians support the alliance with the US, and well over 60 per cent support the idea of US forces being permanently based in the country. What's more, this friendly sentiment is mutual. US service personnel consistently rate Australia as one of the most desirable overseas locations to visit.

There now appears to be scope to negotiate extended deployments of US forces at joint Australia–US facilities. This could ease pressure on American basing in the Indo–Pacific, provide a much firmer, more dispersed, and more resilient allied operating presence and reinforce allied deterrence in the theatre.

The bottom line is that it's time for Australians to consider how best to ensure the country's security in a far more challenging Indo–Pacific region.

Let the debate begin!

Ross Babbage is a Foundation Governor of the Institute for Regional Security, Managing Director of Strategy International (ACT) Pty Ltd and a former senior official in the Australian Department of Defence.

Appendix 4: The Vast Pacific

If there's another war in the Pacific, this is how it will play out.

The hardest of all political facts to change is geography. Faced with the problem of whether it could project sufficient power into the Western Pacific to defeat a major naval power, the United States Navy's War Plan Orange (WPO) before World War 2 secretly concluded that it could not. The inner logic of WPO dictated that the relief of the Philippines would have to await the full mobilisation of American strength. In the meantime, it would have to be conceded to its more powerful neighbors.

The persistence of geography was reaffirmed by an announcement, almost unthinkable to the immediate postwar generation, that Japan was negotiating basing arrangements in the Philippines. The Associated Press announced Manila's intention to begin negotiations to conclude a Visiting Forces Agreement with Tokyo. "A visiting forces agreement would allow for refueling and other logistical and legal needs for periodic visits by Japanese troops, Aquino told a news conference at the end of his four-day visit to Tokyo. The Philippines has similar deals with the U.S. and Australia."

The arrangements between the Philippines and Japan recognise the primacy of geography in two ways. First, it demolishes the illusion, long cherished by the Philippine Left that the US is objectively the primary *potential* threat to Filipino sovereignty. China, by virtue of its size, population, proximity and power occupies that factual position. Second, it concedes that Japan, not the United States, constitutes the major proximate source of reinforcement in the event of a conflict.

The fact is that China is close and Washington is far, far away.

As in 1941, the basic problem in any conflict is whether the US will cross the Pacific in sufficient power to make a difference and if it can do so in time. China's enunciated naval strategy is to take The First Island Chain and dominate the seas all the way out to the Second. The problem for the Navy, and for shipping in general, is that the seas between the two chains will be contested. In fact, controlling this inter-archipelagic space is the key to defending both the Philippines and Japan.

These waters are in time of peace a highway open to the world. Upon them today flows the richest fleet of merchant trading humanity has ever seen; it contains the lifeblood of four of the world's richest nations — China, Japan, South Korea and the United States. Wiser statesmen would have been content to leave things as they are, instead of being tempted by the weakness of an administration in Washington. But be that as it may, in time of war these trade routes will become the arteries on which the survival of great nations will depend. The problem for Chinese adventurism is that making a play for the near seas would bet its entire existence as a modern nation.

Nowhere is the vital importance of these roads clearer than in the flow of oil. A vast stream of tankers comes abreast of the Luzon Strait before going their ways to supply the economies of Japan, China and South Korea. Not just tankers, but container ships headed for largest ports in the world along the China coast and in Japan. If conflict breaks out in the Western Pacific, it's winner-take-all for either China or Japan.

Both Japanese and Chinese naval strategy are dictated by these basic facts. If Japan can hold the First Island Chain it can starve out China. If China can take the First Island Chain, it will starve out Japan. The Japanese counterstrategy to China's is exactly symmetric: control the chain. In Japanese nomenclature, the First and Second Island chains are called the Tokyo-Guam-Taiwan Triangle or TGT. You can see how this works in a map and analysis by the RAND Corporation. First note how the TGT line forms a triangle of undersea mountain ranges which break the Pacific's surface as island chains. It is widely believed that the TGT is heavily wired with sensors. It is certainly patrolled by ASW assets (in which most of Japan's naval investments consist) and possibly by UUVs and other wizard devices.

It will be instantly clear from the map that if Japan can defend the interior of the TGT, it can do two things at once: 1) prevent Chinese SSNs and aircraft from significantly cutting its lifeline to the Americas and the Middle East (though via the north of Australia) and 2) blockading China

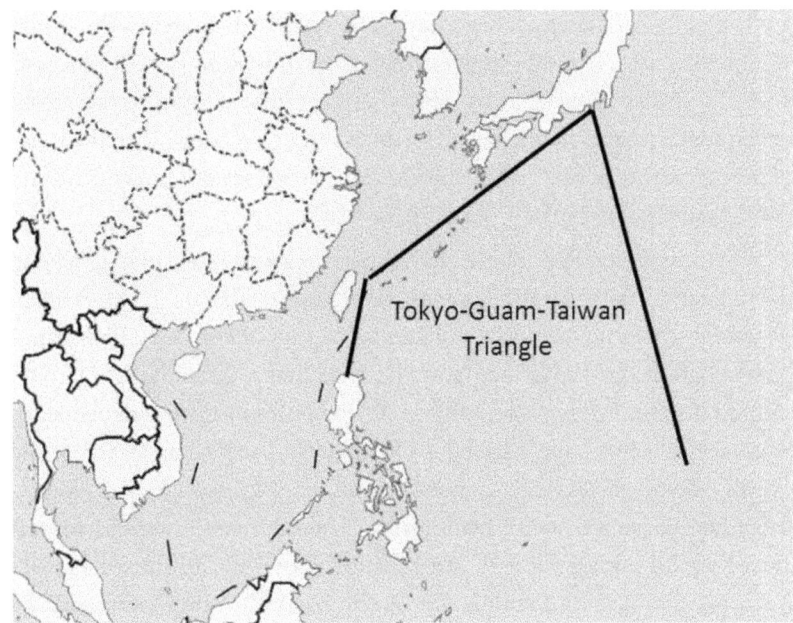

into the bargain. Hold the TGT and the survival of Japan is assured and the near-blockade of China becomes a powerful threat. Lose the line and Japan is probably finished.

From here all the crucial points are easy to understand and can be identified in relation to the map above. China is claiming the Senkaku islands because they lie on the TGT line, between Okinawa and Taiwan. If they can take that, they can move around the south of Okinawa to mine and blockade Japan. In preparation for that they have created the largest mine warfare force in the world.

Defensively the United States is developing Ulugan Bay in Palawan to block access to the Sulu Sea. The Camilo Osias Naval Base on the extreme tip of Luzon has the potential to become crucial. The Sunda and Malacca straits are important chokepoints. And of course there are the bases in Guam, Japan and Okinawa. But the keys to whole defense are Taiwan and the Philippines. Japan has an existential stake in holding these. Without them the whole TGT falls and Japan must reconcile itself to being a vassal of China.

But does Washington have any such stake? Or more to the point, does Obama? The USN arguably has the fallback of distant blockade from the Indian Ocean. It can interdict the Chinese supply route from much further back than Japan. This is an inferior strategy. But the point is that the national survival of the United States does not directly depend on the survival of Japan — or even Australia.

Historically the Republicans have had a greater interest in maintaining a connection with the Philippines than the Democrats. The William McKinley-Leonard Wood sentimental legacy kept the Philippines garrisoned in the 1930s long past the time that a declining USN could defend it. The Tydings-McDuffie Act, whch provided for independence was motivated in large part by a Democratic Party (then based in the South) desire to exclude Filipinos. The act reclassified all Filipinos, including those who were living in the United States, as aliens for the purposes of immigration to America. A quota of 50 immigrants per year was established. Franklin Roosevelt was only barely persuaded by Douglas MacArthur to retake the Philippines in 1944.

Nimitz argued for an advance for Taiwan, bypassing the Philippine Islands, noting that it would place the American military at a better position to strike the Japanese home islands than Luzon. Additionally, Nimitz also argued that by holding Taiwan, the island would serve the political purpose of pleasing the Chinese; militarily, it would also open up a gateway to Sino-American operations on the Chinese coast. Nimitz's recommendation, naturally, placed more emphasis upon the use of the Navy and the Marines, which did not sit well with MacArthur, who already believed that Nimitz was a part of the conspiracy to make the Navy the premier service of the American military. Additionally, MacArthur did not understand how such plan could work. "Just how to neutralize and contain the 300,000 Japanese troops left in [Nimitz's] rear in the Philippines was never clearly explained to me", noted MacArthur. "Admiral Nimitz put forth the Navy plan, but I was sure it was [Ernest] King's and not his own." Roosevelt knew that the plan Nimitz presented had not yet been agreed upon by all US Navy leadership; Raymond Spruance, for example, was not a supporter of the plan for Taiwan, and vouched for an invasion of the Philippine Islands followed by Okinawa, Japan.

Roosevelt, acting "entirely neutral" during the meeting, made no decisions. However, on 9 Aug, MacArthur received a letter from Roosevelt noting that as soon as he returned to Washington, DC, he would push on the plan that MacArthur recommended. It was possible that, like MacArthur suspected, Roosevelt sided with the general so not to upset him and his vast number of conservative supporters during an election year. So it was decided that MacArthur was to personally oversee the landings on the Philippine Islands, assisted by the entire Third Fleet under the command of MacArthur's old friend William Halsey. Some criticised that the entire Philippine Islands campaign was MacArthur's own obsession for fulfilling his personal promise to the Filipino people. In retrospect, although the Philippine Islands made more strategic sense than Taiwan, MacArthur's campaign to clear every corner of the islands of its Japanese occupiers was rather wasteful in time and resources.

Washington's support is by no means assured. It was not in 1944; it is not in 2015. For these reasons one can argue that the only major power whose national interests existentially align with the Philippines is Japan, and to a lesser extent Australia. But Australia is inconsequential as a great power. That leaves only Tokyo. To most Filipinos Japan is an alien nation, while the United States is barely foreign. Yet with Obama in the White House the Philippines must reluctantly realise that if the balloon goes up it will be Tokyo, not the White House, to which it must primarily look.

Geography is invariant in one other thing. Because the Philippines is important for its location, in any conflict it will be ravaged and the casualties to its population will be immense. For most of the Pacific century the Philippines has been, like Poland, in the dubious position of being in the way of great powers going both ways. If the Pax Americana collapses, the Philippines will, as it did on December, 1941, be the first to know.

Richard Fernandez, The Belmont Club

Appendix 5: Ambassador Kim Beazley's Address and Remarks, Perth 13th August 2015

We are actually meeting here at a really interesting point and I think some of the contradictions there are in the Chinese economy and the political consequences of those will become apparent. I think these are going to start to manifest themselves enormously over the next little while and not to our advantage. I think have a different mindset to Americans about the relationship with China. Basically our mindset is around the days when we got $130 per tonne for iron ore. That's gone. It will never happen again. It is down at about $55 now. It is not going to get much more than that. The sort of demand that China has had for resources which has driven the character of our economy for a long time – forget about it. We have to change our mentality on how we handle industry policy and we have to look at where we get our income from as a nation. We have to have a complete rethink of where we stand now. I am afraid that if you have been watching the situation closely the signals are very bleak over the course of the last six months. We used to say well we've got choices. Well we don't have choices. Not on that front now about where our most significant relationship is. That unquestionably, not only militarily but also economically, lies with the US.

When I went to Washington over five years ago, I thought I got the appointment largely because I had a record as an Australian Minister for Defence. And whether the public understands it or not, the relationship we have with the United States in its most important aspect is run by the Department of Defence not by the Department of Foreign Affairs. My Foreign Affairs colleagues would hate to hear me say that but it is true and it is reflected in the staffing of the embassy. There are more people associated with the Department of Defence in the Washington embassy than with the Department of Foreign Affairs and Trade. We spend in US defence industries $10 million a working day. My embassy manages something like 496 foreign military sales contracts.

We are in the middle of the development for the first time, in

Australia's strategic life of a comprehensive air defence of Australia that will be capable of dealing with all comers bar one that decided to use nuclear weapons. We have got that entirely from the United States and that powerful air defence that is being built up at the moment is a product of American genius. Because of the character of the relationship we have with the United States, those systems that we get work at their very best. They work at an American standard. That is not necessarily the case in the outcomes of all American sales to all countries.

So it is still there. The joint facilities with the United States – some of their functions have been rendered technologically obsolete and you can send that in the collapsing down of the number of joint facilities here essentially to those now that are at Pine Gap but at the same time we are developing new joint facilities related to the next generation of surveillance requirements and in Western Australia we are developing a number of facilities related to space surveillance and operations in space. That's the next generation of joint facilities coming through. Stephen Smith had a lot to do with negotiating them so if you want to know the detail, ask him.

That's the character of the relationship and that's what I thought I was about – an understanding of every decision that is taken and its effect on the relationship. It's 31 years since I became Minister for Defence. If you reflect on this fact - the principal surface warship that Australia had when I was Defence Minister was a destroyer – DDGs. And they were American ships and kitted out with American comms and the like. There is more bandwidth now in this (holding his mobile phone) than there was in the DDGs that were in service with the Australian navy at the time I was Defence Minister.

There is a different dimension now that is important. There is a different American attitude, there is a different dimension that it was obvious to me that we need to concentrate on. I will say this briefly and get into more depth in question time. Obama is authentically the first post-Cold War political leader of the United States. He arrived in his office without the Cold War structures and Cold War alliance management which for somebody like myself with an unreconstructed

Cold War mentality which you can't help – once it's in there, you are hardwired. And it is almost impossible to change the perspectives you have on the identification of the hierarchy of important things if you happen to have been, as I was, a defence minister during the Cold War. So it is obvious to me when you suddenly see an American leader come on the scene for whom those points of reference are not significant.

Obama is not so much focussed on the alliance building and arrangements that Americans and us and the Europeans in particular are so used to. He is issues--oriented, global issues-oriented, regional issues-oriented and he seeks his friends from those who can help him. And there more of an equality of friendship in his mind than has been in any of his predecessors - if you are helping me now you are important, if you are not helping me now you are a nuisance. And he is a humble bloke in many ways in the way he conducts US foreign policy. And if you want to do things that attract the favourable attention of the United States, you don't do it any more by stressing what you are, you stress it by what you do. And we do a lot with the United States and in a lot of areas that are important to Obama we are noticed as being there. Nevertheless, his style takes a bit of getting used to. And we are in the process of getting used to that, just as he is about to leave. Anyway, that has been a fascinating aspect of the job. I won't go any further into it now, I am happy to take questions. I'll use the balance of my time here to talk about the economic relationship.

Here's the rub. We have got used to thinking our relationship with the US isn't about the economy, it is about security. You're joking. Australian capitalism is voting with its feet. I'm not talking here about commodity capitalism, I am talking about capitalism producing new services and products. The future of the Australian manufacturing industry lies in the United States and that's all there is to it. That section of Australian capitalism and Australian industry has no voice. It is not reflected in either the Labor Party or the Liberal Party. The Liberal Party tends to be heavily drawn from professional classes and when it comes down to enterprises, essentially suburban enterprise. The Australian Labor Party tends to be drawn from the public sector, the trade union movement, teachers, a particular character of the professional class. Australian capitalists, those who invent or secure the patent for a product that is either part of a supply chain or a niche in community demand, they haven't got time for politics. They run on a straight racer. And they have only time to work out where they are going to get a leg up from.

The United States generates about 70 per cent of the world's risk capital. If you want to get something working, you have to get at least an outlet if not a production facility in the United States. Australian companies in their thousands are marching to the United States. And they do that for many reasons, party for capital and also more importantly for legal protection. You can win in an American court. You struggle to win in the courts of any of our neighbours. I had that demonstrated to me absolutely when I went to the US. A US company had stolen wifi which was a product invented by the CSIRO and patented by an Australian company. It had stolen it, it had not acknowledged it, it had not paid compensation, it was taken to court, it lost in the US Supreme Court nine-nil and was ordered to pay the Australian patent owners a very substantial sum of money.

No other country that we would want to do business in would deal with us in that way. And therefore if you have a certain amount of capital at risk you go where it is safe. People go to the United States and that is all there is to it. No point in arguing the toss on the assumption. And you can see it in the pattern of investment. US investment in Australia in Australian dollar terms is now about $800 billion. That's about ten times I think Chinese investment and that difference, even though China is coming up the list, is growing further apart vis-a-vi the United States. The more interesting story is the other one and that is Australian investment in the United States which now stands at a bit over $550 billion, rising at the rate of $30 billion a year. That's direct and indirect investment. And that reflects heavily of course the fact that we have the fourth largest sum of money under management globally courtesy of our superannuation system and a lot of that is indirect not direct investment because in the United States you have safety, so enormous amounts of your superannuation money is currently being put by money managers in the US but a rising proportion of it is direct investment.

There'd now be close to ten thousand Australian companies operating in the United States, probably a couple of thousand at least with production facilities and as the ambassador I see more and more because they come to talk to me in Washington. Spreading out over the US, pretty intensively concentrated in California where there are about 20,000 Australians in silicon valley, Palo Alto has hundreds of Australian companies. About 45,000 in Los Angeles but most of those are media related. New York has always been strong. There's about 20,000, mostly in finance. On the new technology side, Boston is a growing centre, on the engineering side Texas is a growing centre. We've just opened a consul general-ship in Houston, Texas.

So this is the picture now. This is where we are going. We invest in the United States 25 times what we invest in China. So these are statistics you never conjure with, you never do because our natural orientation is to look to our strategic interest and we have to live in this diverse region and we are right to think strategically in the forefront of our thinking – how we stand in the region around us because that's where we survive. What we need to be is clever and compete. Not to assume that where we ought to focus strategically is where we are situated in practical effect. We've got to become mature thinkers. We've got to able to see absolutely clear-eyed the region around us. And when each of our friends in the region are getting into trouble, we do need to think about that in relation to our friends in China and we need, the Americans need a prosperous China, and that's becoming a very close run thing.

So we are bound up in the success of President Xi and at the moment it's a work in progress, I'm not so sure. And the American president is looking closely at his relationship with China and we also need the capacity for complex and multi-layered thinking about our friends in north Asia, understanding that their engagement with us important.

I think Australia and India are two good nations separated by cricket. In the American conversation with India, which is an intense conversation, the Americans try very hard for Australia, try very hard for Indians to engage Australia effectively because they want to see a strong relationship between friends and potential friends operating in the region. There is amongst the decision-making authorities in India a more complex view prevailing. The Indians are two economies. One is probably the most technologically advanced vibrant economy in the world and it is absolutely the epitome of Indian brilliance and the other is what India once though Britain was, and mistakenly pursued it, and that is atrophied, it is protective, it is inward-looking and would stiff arm the obligations that come with global engagement. The Chinese are essentially mercantilists with a lot of problems to worry about in potentially exploiting relations. China is immensely outward-looking. Chinese economic engagement is immensely outward-looking. Indian private sector engagement is immensely outward-looking. Indian government engagement is immensely inward-looking and deeply suspicious of what others may have invented for them. This is going to take a long time. It is going to be hard to chalk up successes.

Infrastructure is critical. One should always draw the distinction between the announcement and the claim and the likely outcome and the effect of policy. The BRICs look enormously impressive, as an entity, until you look inside them and then you find that their principal point of reference is not to each other at all. In many ways they are bitter competitors. And in many ways the relationship between them is not good at all. And their points of reference economically are in effect with others. The relationship between China and the United States is more important to China than any of their relationships with the BRICs, and the Chinese know it. So the BRICs are a useful pressure point in other parts of the global economy where the Chinese want to have a say. Russia – the Russians are in a lot of trouble. And one cannot accurately predict where they are going, I don't think. But there is one thing you do know which is that when they move out of the resource sector, they struggle enormously.

The Asia Infrastructure Bank is a very interesting point. There would be a lot of people in Washington who think that the administration's handling of it was not well done but at the same time the administration's concerns about the character of that bank were well placed. In my time in Washington I haven't seen a better performance from our government than in their handling of the Asia Infrastructure Bank.

It was a Chinese mechanism to get other countries to pay for the consequences of their overheated construction sector where the Chinese are really struggling now from the devotion of resources to infrastructure that China no longer needed. So they needed an opportunity for their workforce elsewhere. The structure of the bank was basically a Chinese-dominated affair where the Chinese controlled the board and then Chinese determination of contracts which were taken from successful applicants for loans would be performed exclusively by Chinese firms. It was an arrangement where they invited Australia to put in $1 billion plus $4 billion to subsidise the consequences of Chinese overcapacity in infrastructure capability. Not good. And so what we did, in consultation with the Americans because the Americans were happy we were moving towards membership, we said we are going to stand for decent governance and integrity in this bank. And almost alone, almost alone, because it was the Australian government, because they wanted a respectable Asian country to deal with, and Australia was important to them in that regard, the countries of Southeast Asia, who felt they had to automatically join the bank put enormous pressure on us to come in and play a constructive role and we spent a considerable amount of time with the Americans discussing what we were discussing with the Chinese. And basically about 85 per cent of the paper we put in to get a decent set of governance arrangements, shareholder arrangements, a situation where the Chinese could not dominate the board without another couple of members of the bank coming in behind them. We have changed what was, I think, quite a cynical ploy, to something that is really going to be quite welcome and we ought to be pretty proud of it.

Prime Minister Abe is taking Japan down a different road. Japan is changing its objective from being a normal country to being a normal ally. If Prime Minister Abe gets his program in place, and there is not an effective revolt against it within the LDP or in the Japanese polity more generally, Japan will have a different aspect in global and regional politics. It will be a substantial player not simply in economic and diplomatic terms but in military terms as well.

If the current trajectory of our LNG deliveries to Japan plays out, I have seen a lot of assessments that Japan will resume its previous status as our principal trading partner. So that's the direction of things at the moment, both in Australia's relations to China and Australia's relations to Japan.

In some respects, Japan is showing signs of a substantial comeback. In military terms we are all much more aware of Japanese capabilities, and they have some brilliant technology, well in advance of both the US and us. And they can bring to the table, not least in submarines by the way.

We don't comprehend the depth of the relationship between the United States and China. Just before we came away here about a month or two ago, they had the annual SME discussions with four hundred senior Chinese officials operating in the United States, 127 outcomes from it that they were working on, trying to get in place a sort of investment treaty. The Chinese really do stiff arm anyone else coming into their economy, including us, and nevertheless want access to everyone else's economy on a non-discriminatory basis. If we gave access on the basis of our access to China, they would hardly be any Chinese investment at all. But the US is engaged in this really deep discussion with China.

You've got to remember that for a hundred years the principal goal of American policy in the Asia-Pacific region has had as its centre a prosperous and united China. That has been the principal American goal in Asia-Pacific affairs. It is why they missed the war. In World War 2 in the Pacific evolved around American attempts to secure a decent defence for China and the Americans finally lit on a sanction which gave the Japanese a choice - would we go under or would we try to force the United States to change its mind? Hence Pearl Harbor. We all tend to forget that. Basically Pearl Harbour was a product of the American defence of China. Now you talk about a gung ho marine, alright, where does 'gung ho' come from? Well it came from the Eighth Route Army, the Chinese Eighth Route Army which was advised by American marines. And so they jog-trotted, they went "gung ho, gung ho, gung ho", which means "work together, work together, work together".

Then you have China as one of the P5 on the United Nations. Why on earth would China be in the P5 out of World War 2? Well, America. America insisted that the rest of the world respect China. Things went pear-shaped after the Korean War but Nixon restored the normal trajectory of the American relationship with China. And the Americans built that into their conversation with Deng Xiaoping and Deng Xiaoping's response to the Americans was to say that all these issues, these territorial issues we will put on the back burner, sign an agreement with the Russians and say that we will one day come for the territories you stole from us in the 19th century but formally at this point of time recognise that we won't, and that basically for the next hundred years or so respect the boundaries as they now exist. We said to the Americans, we have a Chinese position in relation to the region around us but we are going to hide that under a bushel, we are going to focus on the growth of China.

Something went wrong on the way to the theatre, and all of a sudden the long term agenda was dragged into the current and has created a substantial problem for the United States who will not permit a derogation of sovereignty of the Southeast Asian nations lying down. They won't. America stands for freedom and they will not permit, even though it's a hell of a nuisance with respect to their overall objectives with China, so they will come into a diplomatic conflict with China and there is a risk that this could blow into something else and that is a substantial preoccupation of American statesmen.

The Americans aren't like us really. They are sophisticated and in-depth so they can sustain multilayered characters of relationships where they can simultaneously try to engage to try to make progress on important fronts and at the same time tell the other fellow "No". The very fact that we begin to ask ourselves the question, do we have to choose between our ally and our principal trading partner is a reflection of Australian immaturity. That we think that that's a question that needs to be asked, let alone become preoccupied with. It is like asking the Americans do you think you should have a relationship with China or only a relationship with Japan. They would say "Say what? What are you talking about?" But that's us – we are so full of ourselves. We regard ourselves as so intelligent in all these areas. We cannot see the wood for the trees and we think we know something about China, we know nothing. Compared to what the Americans know and what the Americans have been committed to. They gave their whole selves to the defence of China. By the way, they didn't do it for Europe. They helped out on Europe. They pushed the Japanese into a decision and given the military character of the Japanese government at the time, there was only decision that Japan could make.

Kim Beazley is the Australian Ambassador to the United States. He spoke at an event conducted by the Perth USAsia Centre on 13th August, 2015.

Appendix 6: Step by Step, Here's How to Fight China

The air campaign in Desert Storm (First Gulf War, 17th January, 1991 – 28th February, 1991) was a watershed for air power. It demonstrated the effectiveness of precision munitions, marked a high water point for electronic warfare and introduced radar stealth in a decisive manner. It also established a template for the application of air power that has taken root in Air Force culture and remains firmly established a quarter century later.

Unfortunately, the Instant Thunder air campaign (the air component of Desert Storm) has also become the template for future air campaigns, despite being poorly suited for that role. In retrospect, we have learned many of the wrong lessons from Desert Storm, in that we had time to build up forces, operated from a broad network of U.S.-built bases and essentially ravaged the military structure of a small, isolated nation with an incompetently led military using obsolete equipment and outdated employment doctrine.

By the time Operation Allied Force (the NATO bombing of Yugoslavia) rolled around in 1999, it should have become clear that the same template produced uneven results at best, even when backed by a combined NATO air force.

In the aftermath of a series of wars against relatively weak adversaries, planning for a larger war has descended into nonspecific terms. Pentagon discussions on force structure, posture and capabilities are often based on a "capabilities-based," generic adversary reduced to the status of an opposition force. This adversary might be referred to as a "near peer," characterized largely by the technology it brings to the fight rather than understood as a living, adaptable enemy that might have to be fought under unfavourable conditions.

This habit ignores the reality that the People's Republic of China has eclipsed the old Soviet Union and its successor as a superpower, militarily, economically, politically and technologically. We remain wedded to an inappropriate warfighting model leftover from the Gulf War, while ignoring China's evolution as a military power.

We ignore this evolution at our peril.

To attempt to apply the Desert Storm air campaign model to other nations is of questionable utility, and applied to China in particular is pure folly. China is large, resilient, can mass military forces like few other nations and is clearly a superior power when fighting in its own territory. Moreover, it has spent a quarter century of military development ensuring that the United States can never be in a position to repeat Desert Storm against the People's Republic.

Chinese military force design has been built specifically to counter the U.S. Air Force's reliance on stealth and forward basing, and to reduce the threat of carrier aviation by developing weapons designed to keep the carriers far away from the action. Our response has been to plan to fight symmetrically, matching our technological widgets against theirs in a battle in the PRC's front yard.

Strategically, this methodology replays the successful strategic campaign, whereby the USSR spent itself into collapse trying to match American technological prowess. This time, however, the United States is on the wrong side of that strategy.

There is benefit of adopting an asymmetric offset strategy to deal with the PRC's general technological parity and commanding position. There is additional benefit of adopting a strategy that could be executed today, without being dependent on technologies that have yet to emerge. The reality of the Chinese force structure is that it is largely a defensive structure whose utility wanes rapidly with distance from the Chinese coastline.

Unlike Imperial Japan, China lacks a carrier-capable, blue-water navy with which to challenge the United States, and has not begun an overt territorial expansion that provides overseas basing facilities. Like Imperial Japan, China is heavily dependent on overseas supply lines, and thus subject to interdiction of critical warfighting resources, especially energy.

China's import dependency is particularly acute for energy supplies, which have to travel long distances through unfavourable maritime terrain, only to then be dependent on a limited domestic transportation infrastructure which is itself energy-intensive. This means that the PRC is vulnerable to a counter-logistics campaign intended to limit China's energy supplies in a fashion that reduces or eliminates their capability to project military power.

The foundation for a military campaign against the People's Republic of China, presumably with the objective of stopping or reversing Chinese aggression, could be based on strategic interdiction, a.k.a. SI — a joint effort designed to prevent the movement of resources related to military forces or operations. An SI campaign would be designed to repeat the fundamental success of the Pacific War — isolating Japan to the point where it could no longer impede Allied operations in the Pacific.

Historical background

A counter-logistics campaign has historical precedent in the Pacific. Indeed, we have volumes of data documenting the execution and effect of such a strategy against Japan.

In February of 1942, Japanese forces wrested Rabaul, New Britain, from the outnumbered and unsupported Australian detachment. In short order, Rabaul became the primary forward base in the South Pacific and a major obstacle sitting squarely between both Allied theaters in the Pacific. Gen. Douglas MacArthur's plan to recapture the island fell afoul of resource constraints and the higher priority held by the war against Germany.

By August of 1943, the President made the decision that Rabaul would instead by bypassed rather than seized, largely because of the emerging realization that Rabaul did not have to be captured in order to be neutralized. Operation Cartwheel, starting in December, neutralized the island citadel without a direct and costly amphibious assault, and without requiring resources above what was already allocated for the theater.

Rabaul was attacked by air, isolated by manoeuvre and starved by air and naval forces to the point where it could no longer be used as a venue for power projection. Australian forces liberated Rabaul without a shot fired, surrendering four days after the surrender ceremony in Tokyo Bay.

While directed against only a small island group, the isolation of Rabaul is a relevant historical example of the success of a long-term strategy to neutralize powerful military forces in a critical position. Operation Cartwheel was a small example of what became a general strategy for the conduct of the Pacific War — that Japanese garrisons would be isolated and cut off, attacked in place and that the home islands would be deprived of materials, energy and supplies that relied on water or rail transport.

By the end of the war, a coherent maritime interdiction campaign brought the Japanese home islands to the brink of surrender, while an air campaign against Japanese railroads tied up domestic transport to the point that needed resources could not even be moved internally.

A well-designed, pre-planned strategic interdiction campaign provides a potential way forward for a war-winning air and naval power application, specifically tailored to the People's Republic of China's specific characteristics. In particular, the campaign is intended to apply lessons learned against Japan to China, as if China were in fact an island.

From a transportation standpoint, China is over 98 percent island. China's international land transportation networks, even in combination, are dwarfed by any of China's larger ports taken singly, and its land transportation already suffers from a lack of capacity and susceptibility to disruption — both exploitable vulnerabilities.

A strategic interdiction campaign is a strategy based on denying logistical supplies to the fighting forces of an adversary. It is a combination of several efforts, including a limited blockade, interference with transportation networks and disabling some energy production at the resource level. The primary objective here is to effectively neutralize certain elements of PRC military power by starving it of energy.

In contrast with maritime interdiction, strategic interdiction is not an airtight blockade but a targeted effort to interdict primarily the production and transport of energy resources all the way back to the source. A campaign would have four elements:

A "counterforce" effort designed to attrit the adversary air forces (particularly bombers), naval forces (gray hulls) and naval auxiliaries (replenishment) to the point where they can neither project military power nor defend against U.S. power projection, at least far beyond the PRC continental shelf.

An "inshore" element, which consists of operations to deny effective use of home waters, including rivers and coastal waters. Standoff or covert aerial mining is a key component of this element.

An "infrastructure degradation" plan intended to disrupt or destroy specific soft targets, such as oil terminals, oil refineries, pipelines and railway chokepoints such as tunnels and bridges. Many of these targets would be in airspace not defended by ground-based air defense.

A "distant" maritime strategy, which occurs out of effective adversary military reach, intended to interdict energy supplies. This strategy is aimed primarily at bulk petroleum carriers (tankers) and secondarily at coal transports, and not at container, dry bulk or passenger vessels. Such a strategy might not be lethally oriented, directed instead towards the seizure and internment of People's Republic of China -bound vessels.

In effect, this strategy targets its effects on naval and air forces, which rely on jet fuel, and leaves the gasoline and diesel-dependent army shore-bound. Along the way, secondary effects ripple through the industrial, refining, power generation and transportation sectors of the economy, with broad effects that are difficult to predict or quantify. A strategic interdiction strategy is not a short war strategy. It is a prolonged containment strategy derived from previous experience in the Pacific War.

While we don't think of the PRC as an island nation, effectively it is one. Over 98 percent of the PRC's external commerce by tonnage moved is seaborne. The transportation infrastructure over land borders accounts for a miniscule portion of the People's Republic of China's imports, and all goods crossing the borders are a long way from China's industrial sector. The total volume of goods moved overland via train, road and watercraft through the borders in a year is exceeded by the port of Shanghai in 60 days, with room to spare.

This reality is effectively impossible to change or mitigate in any significant way, and clearly indicates the potential of a Strategic Interdiction campaign focused on maritime transport.

Energy — the sixth ring

The targeting strategy for the Gulf War's air power application was based on Col. John Warden's "five rings," which threatened the subject country (in this case, Iraq) as a series of concentric rings. The outermost ring (fielded forces) protected the inner rings (population, infrastructure, organic essentials and leadership). As the theory went, one of the key advantages of air power was that aircraft could fly over the outermost rings to get to the key one — leadership.

While applicable to Iraq in 1990, the applicability to China is questionable, as it is not a centralized Ba'ath Party dictatorship led by a single individual. Furthermore, it is risky to attempt to execute a decapitation strategy against a state with a significant nuclear arsenal. Instead, a Strategic Interdiction strategy is centered on the sixth ring, which doesn't exist at all in Warden's construct except as part of the second and third rings.

The Sixth Ring is the energy ring, which also serves as the glue that holds all of the rings together. In this modified construct, the center ring is still a physical target, but under an Strategic Interdiction strategy, it is not one that is attacked directly. Effects aimed at it, along with every other ring, are secondary effects of an energy denial strategy.

China is a massive energy consumer, relying primarily on coal for electricity and oil for transportation. The two are not really interchangeable, and each has its own vulnerabilities. Coal-fired power plants provide approximately 70 percent of China's electricity generation, a percentage that has remained relatively constant since 1980. Nuclear, natural gas, solar and hydropower are a comparatively small portion of the power generation infrastructure, providing less energy combined than oil does alone. As these last four are comparatively minor energy sources, they are ignored in this analysis.

Coal

China is the world's largest coal consumer. Steam coal is used for power, and coking coal for industrial processes. Coal consumption is largely taken up by industry, including power generation. Even without counting heating demand, the power sector consumes more steam coal than industry.

China produces most of its coal domestically, producing 3.87 billion tonnes of coal in 2014 and importing another 291 million tonnes in 2014, a domestic/import ratio of better than 13:1. In the past two years, Mongolia has emerged as a key supplier of imported coal, supplying by train and truck rather than by ship.

In 2012, China had 58 coal offload ports, scattered all along the coast, serving both domestic and international coal movement.

While imported coal appears to be a drop in the bucket compared to the total coal supply, this is not true for all regions. Seventy percent of imported steam coal was consumed by power plants in coastal regions south of the Yangtze (Guangdong, Shanghai, Guangxi, Zhejiang and Jiangsu) — the demand centers furthest from China's main coal-producing regions. This may not be related to the capacity of the transport system but its cost — for the south-eastern provinces it is cheaper to import coal than to ship it domestically.

Oil

Crude oil accounted for roughly 19 percent of China's electricity consumption in 2012, making it a distant second to coal. Oil supplies are mostly gobbled up in transportation, although diesel is also the fuel of choice for backup power generation. China's appetite for crude is massive, requiring imports of 2.26 billion barrels and another 219 million barrels of refined fuels on top of domestic oil production of 1.53 billion barrels in 2014. In total, in 2014 China imported 56 percent of its oil needs.

The lion's share of petroleum consumption is taken up by industry, including electricity production, chemical manufacture and refining. The transportation sector in China consumes almost as much petroleum as industry, consuming the vast majority of middle and light distillates burned in a year. Transport accounts for 46 percent of the gasoline consumed, 91 percent of the kerosene and 63 percent of the diesel fuel.

Oddly enough, as much as two thirds of China's annual diesel fuel consumption is burned transporting coal. By comparison, the entire transportation sector consumes less than two percent of the electricity used in a year.

China is making an effort to establish a strategic petroleum reserve (SPR) for crude oil. In 2010, China had a commercial storage capacity of between 170 and 310 million barrels, but no national strategic reserve at all. The People's Republic of China's tenth five-year plan (2000 to 2005) marked the beginning of the government Strategic Petroleum Reserve program. Phase 1 established a capacity of 103 million barrels at four sites and was filled by 2009; phase 2 is expanding that to by another 226 million barrels at nine locations, of which 210 million barrels will be filled by the end of 2015. The last phase, (2020), should bring the Strategic Petroleum Reserve capacity to half a billion barrels of crude oil.

Even at this capacity, the Strategic Petroleum Reserve holds less oil than the People's Republic of China imports in three months. The Strategic Petroleum Reserves holds no refined products, which are entirely reliant on a commercial storage capacity estimated at 400 to 480 million barrels for all types of refined fuel combined. With one notable exception using a reclaimed salt mine, the SPR sites are conventional above-ground storage tanks, often on the coast, and often next to existing refineries.

Internal transportation network

China has a well-developed transportation network all along the eastern corridor, consisting of waterways, roads and railways. Compared to the United States, China's water transport enterprise is massive while the pipeline transport infrastructure is minuscule. As of 2013, the Chinese rail network consisted of 90,000 kilometres of conventional railway lines and another 10,000 kilometres of high-speed lines, which are mostly passenger lines. Of this, 56,000 kilometres was electric and 48,000 kilometres double-tracked.

The country has 125,000 kilometres of navigable inland waterways, including the Yangtze River, which moves more freight by far than any other inland waterway in the world. The public road network consists of 4.36 million kilometres of roads, 34 percent of which are dirt with 424,000 kilometres of highways including 9,600 kilometres of expressway. In 2012, the country reported having 9,100 kilometres of oil and gas pipelines, roughly 0.3 percent of the U.S. pipeline infrastructure.

The transportation network is substantially less dense away from the eastern provinces, and is comparatively sparse at the country's borders or in the west. With respect to the tonnage of freight moved (which includes fossil fuels), China uses highways, waterways and rail, in that order, to move goods internally.

Air transport is virtually insignificant by comparison, while pipeline transport for oil, refined products and gas is comparatively limited. Measured by tonne-kilometres rather than simply tonnage, waterways and highways switch places, because waterways are used to ship goods longer distances by far. In 2012, the average tonne of freight moved 1781 kilometres by waterway, 748 kilometres by road and a mere 187 by road.

Many trips mix modes of surface transportation. The implication of this transport distribution is that China's internal transport is reliant on the two modes that are most oil-intensive. In 2014 total freight traffic increased by over seven percent compared to 2013, with roads and waterways gaining traffic (10 and 16 percent increases, respectively) and rail losing it (5.6 percent decrease).

It takes energy to move energy. Coal accounts for a full 52 percent of the tonnage shipped and 40 percent of the tonne-kilometres hauled by rail and 21 percent of the domestic freight handled in the large coastal and river ports. Petroleum products account for only four percent of the rail tonnage and nine percent of the port freight. On average, a tonne of coal moved by rail travels 647 kilometres.

Moving coal is nontrivial in China. The three top coal-producing provinces are Shanxi, Shaanxi and Inner Mongolia (outlined in black), which alone account for more than half of the national coal output. These three provinces are some distance from the coal-consuming provinces.

The railway network was unable to keep up with the transport demand as China's coal usage increased, and as a result from 1997 much coal traffic was diverted to multimode transport, where coal is carried by rail to the ports on the Bohai Sea and thence by coastal shipping to the south. Truck transport is used extensively, resulting in world-class traffic jams. In 2010, Inner Mongolia coal traffic generated several major traffic jams, extending for more than 100 kilometres and lasting for days.

The difficulties moving coal often forces provinces far from the producing regions to ration power consumption in response to supply disruptions, including inclement weather. The strained coal transportation system is already imposing local coal shortages on the power industry, with the impact greatest on the south-western provinces (Tibet, Sichuan, Chongqing, Gansu) and the provinces south of the Yangtze. Oddly enough, Shanxi province exported so much of its production in 2012 that its own power plants ran short.

Refining sector

Crude oil cannot be burned for any purpose until it has been refined. In short, getting refined petroleum products is dependent on the quality of the oil that goes in and the equipment available for processing the oil. Some products are distilled, while others are chemically broken down and reformed. Oil is full of impurities, especially water, salt and sulphur, which must be removed during refining. Chinese oil imports are largely Middle Eastern, heavy "sour" oils which require more refinery processing than the "light, sweet" crude produced elsewhere.

The fuel that is most important from a military power projection standpoint is jet fuel, a high-quality mixture of kerosene, naphtha and additives used by aircraft and turbine-powered ships. Without fuel, aircraft are grounded and warships remain in port. One of the goals of an Strategic Interdiction campaign it to make it really hard or impossible to make jet fuel. Turbine powered ships can operate with marine diesel fuel (the U.S. Navy runs ship turbines on it) but aircraft turbines cannot.

In the past decade, the People's Republic of China has undertaken an ambitious effort to increase its refining capability from six million barrels per day in 2000 to 12.6 million barrels per day in 2013, while simultaneously consolidating into fewer refineries of much greater size. As a result, there is excess capacity remaining and the number of lucrative targets has been reduced and refinery functions consolidated. The refinery sector operated at only 81 percent of capacity in 2012, which has turned out to be a mixed blessing.

This excess capacity actually delayed further expansion of domestic refineries originally planned for 2016 and 2017, leaving the Sino-Burmese pipeline unable to deliver oil for refining because the ground has not been broken for the refinery site that would have received the imported crude.

As late as 2012, China did not meet all of its refined fuel requirements with domestic refining, and in 2012 one out of every four barrels of petroleum imported was actually a refined product. As the market shifted, so did the mix of refined fuels, as producers chased the more profitable products, especially jet fuel. In 2014, China was a net exporter of all refined fuel products except naphtha. This occurred despite the fact that China's surviving smaller "teakettle" refineries, which account for a quarter of the nation's refinery capacity, produce no jet fuel components at all.

Like coal, China's refinery infrastructure is not evenly distributed.

Refinery capacity is concentrated in the east, with a scattering of refineries along the sole railway link to the far west. Refineries in the country's interior are largely reliant on domestic feedstock. Teakettle, or small privately-owned refineries, have to acquire a permit to use imported oil at all. Critically, the refineries along the coast are more reliant on imported oil, and the four southern provinces are close to 100 percent reliant on overseas imports for their feedstock.

Strategic interdiction

Given China's unique energy vulnerabilities, combining massive demand, significant imports and a capacity-challenged transportation network, a military campaign designed to apply pressure at multiple points in the energy web would seem to be both cost-effective to execute and difficult to counter, even under conditions where operations in the Western Pacific are limited in scope and duration.

The objectives of such a campaign would be to so disrupt the energy and transport sectors of the People's Republic of China such that there is a pervasive and enduring effect on fielded forces. The campaign design takes lessons learned from the Pacific War against Japan, where both the Imperial Japanese Fleet and its air arm were systematically deprived of fuel, which affected all aspects of their military enterprise from engine testing and training to flight time and vital resupply.

A strategic interdiction campaign rests on four pillars and is intended to provide a viable offset strategy that is based on a presumed need to coerce a specific adversary in a designated region — China in the Western Pacific. The campaign is a long-term, counter-logistics effort which rests on four pillars: counterforce, inshore, infrastructure degradation and distant interdiction.

I. Counterforce

The counterforce pillar is intended to neutralize any PLAN (People's Liberation Army Navy) or PLANAF (People's Liberation Army Navy Air Force) attempt to project power outside Chinese coastal regions and is built in expectation that the PLAN and PLAAF (People's Liberation Army Air Force) will come out to fight. In fact, such an adventure against Taiwan, the Senkaku Islands or any number of island possessions may be the event that requires a U.S. response in the first place. The PLAN may conduct an amphibious operation, undertake convoy escort or execute any of the out-of-area missions that a blue-water navy would aspire to.

It may be desirable to sink surface combatants, but also replenishment ships, auxiliaries or minesweeping vessels. It is also permissible to attack blockade runners regardless of ownership, an issue of particular importance to the fourth pillar.

PLAAF bomber aircraft armed with cruise missiles will undertake counter-maritime and counter-land missions at some distance, perhaps as far as Guam. It will be necessary to counter these operations, often from a standoff position. In the Pacific, the long expanses of open ocean will require a focus on counter-air and counter-maritime capabilities. U.S. anti-ship capabilities have long since been allowed to atrophy, even in the Navy, as the PLAN has fielded increasingly capable anti-air-warfare ships which must be attacked from increasingly long distances.

Without diving into specific weapon and sensor combinations, standoff and specificity are key anti-ship weapons attributes, and any aircraft or vessel that launches them must have a suitable sensor system or a connection to one.

The simplest method, and the most difficult to affect by enemy action, is for the launching unit to have its own system for detection, identification and targeting of its on-board weapons. This is already approach used by fighter aircraft for air-to-air targets, and by all surface combatants. This approach could be extended to include counter-maritime capabilities.

Improved long-range sensors, especially radar and ELINT sensors useful in anti-surface warfare, could transform our bomber fleet into the transoceanic counter-maritime force that it used to be. Increasing the effectiveness of counter-air capabilities is also a key component of this pillar.

II. Inshore

Inshore operations are closely related to the counterforce pillar; there is significant overlap in capabilities. The purpose of inshore operations is somewhat different — the inshore pillar is intended to deny the PRC the unfettered use of waterways, rivers, harbors and offloading and replenishment facilities.

The objective is twofold; to prevent the PLAN from being able to sortie, sustain at sea, and reload or replenish, while simultaneously interdicting energy supplies which are transported by oceangoing, coastal or riverine vessels. Strictly speaking, with the exception of river mining, this pillar does not require direct attack against the mainland, and relies as much on the threat of attack as actual attack.

Aerial or covert mining is a significant component of the inshore strategy, capitalizing on both the effects of actual mines and the suppressive nature that fear of mines has on shipping. Aerial mining is the only way to lay large offensive minefields quickly, while covert (underwater) mining may allow for precise placement of advanced mines.

The Yangtze was mined by USAAF (United States Army Air Force) in World War II, and the Rangoon River in Burma was entirely closed to Japanese shipping by aerial mines. PACOM (United States Pacific Command) has recently demonstrated the Quickstrike-ER, a standoff, precision version of the legacy Quickstrike bottom mine. Combined with the shorter-range Quickstrike-J, the U.S. is now developing the capability for one aircraft to lay a minefield in a single pass.

Combined with underwater minelaying, low altitude insertion or stealth aircraft, there is an emerging capability to lay minefields in areas where it was previously infeasible, including rivers, river mouths, and harbors. Smart target detection devices allow both limited selectivity of targeting and resistance to minesweeping.

The inshore pillar is aimed primarily against the waterborne element of the transportation network, with secondary effects against naval facilities and ships. It is intended to apply against domestic, short-haul shipping, and against ships carrying critical imports which penetrate an allied naval cordon. It would be possible to interdict vessels at either end of the network for domestic traffic — coal traffic might be bottled either at the on-load or offload facilities. Fear of mines may be more effective at halting traffic than actual mines themselves. While under the 1907 Hague Convention all minefields have to be declared, not all declared fields have to be mined.

In many cases, once mines have been employed somewhere, they could have been employed anywhere and this uncertainty is a powerful deterrent to movement.

III. Infrastructure degradation

Interdiction of maritime transport alone will not necessarily achieve the full goals of the campaign by itself, although it will likely have a devastating (though reversible) effect on People's Republic of China's industry and power generation. The People's Republic of China's domestic energy supplies, combined with refining capability, ensure that the military could still be supplied with sufficient energy supplies to conduct sustained operations, albeit at a significant cost to other domestic priorities.

Local energy shortages will likely be exacerbated and reallocation of suddenly scarce resources would be challenging even for a country where the actual flows of resources were well known. The infrastructure degradation campaign is intended to give the resource denial efforts a push in the wrong direction by disrupting, incapacitating or destroying critical chokepoints in energy transport and production.

The most lucrative targets are rail tunnels and bridges, certain refinery components, international oil pipelines and oil transfer terminals. Nonlethal means may be used in addition to lethal ones, although even a nonlethal attack on petroleum handling or refining facilities can result in a lethal catastrophic effect.

The infrastructure degradation pillar is intended to constrain overland imports, while simultaneously destroying the refinery capacity necessary to turn strategic reserve or domestic crude oil into usable fuel and interdicting rail and water transportation at their most vulnerable points.

IV. Distant interdiction

The distant interdiction pillar involves a maritime interdiction effort aimed specifically against ships bound for China with energy cargoes, particularly oil, refined oil products and coal.

It is the most legally complex of the pillars in that it involves action against both Chinese and foreign-owned shipping. It is also the pillar that can and should consist largely of actions that involve minimal property destruction, although it does involve the use of force. It takes advantage of the fact that the vast majority of China's imported energy supplies come through chokepoints that can be easily interdicted. The distant interdiction effort stretches from the Asian continental shelf all the way back to the original points of embarkation.

The maritime geography is unfavourable for China. Unlike the United States, which has four coasts that are mostly devoid of potentially hostile neighbours (excepting Cuba, of course), China is hemmed in by island chains that are owned by nation-states with longstanding territorial disputes with China. Supply lines across the Pacific from the Panama Canal or South America pass nearby U.S. territory on the way.

Furthermore, China has neither a true blue-water navy nor a robust network of forward bases, and cannot project naval power long distances from the mainland. In short, the People's Liberation Army Navy cannot protect its supply lines for energy back to the sources, which are typically in the Middle East for oil, or Australia for coal.

The distant interdiction portion of the campaign would aim to define energy supplies as contraband and to intercept, board and intern vessels carrying energy supplies to China. This would include vessels that are Chinese-flagged and foreign-flagged ships carrying energy to China. The vast majority of ships, which are container ships, are of no interest and can be allowed through, but petroleum tankers (oil, oil products and LPG) and bulk coal carriers would be boarded, seized and interned. The nature of these ship designs makes them the easiest to identify and greatly simplifies the execution of a blockade.

Under threat of attack, neutral ships may elect to avoid the conflict area, carrying other cargoes to other ports. There is little profit in attempting to deliver bulk cargo while risking damage or loss of the ship. Under such conditions insurance rates typically rise, and the premium for a brief exposure may reach upwards of 10 per cent the market value of the vessel, plus cargo value. The internment of Chinese-flagged vessels or neutrals with contraband bound for China is a compound-interest challenge.

Every internment not only removes the current cargo from the delivery sequence, but removes all subsequent cargoes that might have been carried by that ship. In the case of very large crude carriers (VLCCs), that can account for very large cargoes indeed. At this time, there are less than 100 Chinese-flagged VLCCs, accounting for under a sixth of the worldwide VLCC stock. Given the favorable geography, the U.S. Navy would not have to spread out far in order to interdict these ships, and may even block chokepoints outside Asia, like the Bab El Mandeb or Strait of Hormuz.

In 2014 an average of around 11 to 15 VLCCs transited the Straits of Malacca on any given date, traveling in both directions. Not all of these were bound for China, and a tanker may in fact carry oil for several destinations on a single voyage. A naval task force, supported by air, could intercept a significant number of these ships and interrupt their transit, either loaded or during the return voyage. Each ship that delivers cargo to China is subject to seizure on the return, providing two seizure opportunities on a single voyage.

Sample targets were compiled for this analysis. The largest target category is rail lines, which are broken at tunnel entrances and bridges to make repair time consuming and difficult. There are 32 targets chosen (white targets) to interdict coal transport (mostly exiting Shanxi and Shaanxi provinces) and international coal and oil imports.

All of the rail transport from these two coal-producing provinces plus Inner Mongolia is interdicted, blocking movement of 70 percent of the country's domestic coal. All railway border crossings were interdicted on the Chinese side. Thirty-two additional rail targets (yellow targets) were selected to shatter the rail transportation network, mostly at river crossings, which are intended to have a secondary effect of blocking shipping channels.

Every railroad bridge along the Yangtze 500 nautical miles upstream from Shanghai is on the list. Combined with additional railroad bridges across other waterways, the rail links between north and south China are severed, excepting only the high-speed passenger lines which are only broken at the Yangtze. Every one of the country's top ten freight corridors is broken in at least one place. Road bridges were only targeted across the Yangtze River (to block ship traffic) or when roads and railroads shared a bridge. Road tunnels were targeted only if adjacent to rail tunnel targets.

Pipelines accounted for six targets (orange), inside China's borders, usually by targeting pumping stations but also the pipeline itself. There are 32 refinery targets (red), all allocated to refineries producing jet fuel, kerosene, and/or adjacent to strategic petroleum reserves. Distillation towers, rail terminals, rail access, power plants, and pumping stations consisted of the majority of aimpoints, with two to 10 aimpoints per refinery.

Water terminals were left alone unless directly attached to a refinery. Some refineries were isolated by cutting the rail approaches at bridges and otherwise leaving the refinery alone. Strategic Petroleum Reserve sites were targeted when adjacent to refineries but not if otherwise located.

There are 39 inshore targets, all minefields (blue). Those minefields accounted for all PLAN bases and all large oil terminals, plus the mouths of the Yangtze and Pearl rivers. No river mining was conducted upstream of any river mouth. Only two minefields are offshore, both at oil terminals in the South China Sea, all others were within the 12 mile limit and often within the three mile limit. Because of the uncertainty involved with mining in defended airspace, most coastal refineries were double or triple-tapped, in that their rail links and refining capacity was directly attacked in addition to mining. Mined oil terminals are essentially double-tapped with the distant interdiction pillar.

No military facilities were directly targeted, nor were communications, underground petroleum storage, air defenses, commercial power plants, coal load/offload facilities, space control, space launch or leadership targets.

The direct effect of an Strategic Interdiction strategy on the People's Republic of China's power projection capabilities cannot be precisely predicted from the data available from open sources. The goal of depriving PLAN and PLAAF forces of jet fuel will not be accomplished within a few weeks.

While China has no strategic reserve for refined petroleum products, it does have commercial storage, plus (presumably) military storage of undetermined size and composition. Diversion from civilian use and reallocation of refinery resources are probable, but both of those efforts will be hampered by interference with transportation; reallocation of production may be prevented by damage to refineries.

A detailed analysis of the anticipated effects is both beyond the scope of this white paper and not suitable for public dissemination in any case. What is certain is that an energy denial strategy will have immediate effects on the People's Republic of China. Interdiction of oil imports will force both an immediate reallocation of resources and likely cause a dip into the strategic reserve. A reduction of coal imports will have a rapid effect on power generation, although a reduction in industrial power use could mitigate the effects of power shortages.

Any perturbations, including physical damage, against the rail transportation system will ripple through the country – the system is over capacity as it is and even weather events disrupt rail transport. Damage to refineries simply cannot be mitigated rapidly — these are the softest of soft targets and even relatively minor damage can cause a refinery to shut down.

It is equally certain that interdiction of coal and oil imports will have a disproportionate effect on the provinces bordering the South China Sea. Aside from the inevitable electricity shortages, oil interdiction will idle every refinery in the four south-eastern provinces, taking 20 per cent of the country's total refinery capacity offline without any need to damage those refineries.

From an interdiction standpoint, it is easiest to interrupt foreign flows, whether they flow by sea or by pipeline. For coal, overseas interdiction is nevertheless worth the effort because of the disproportionate impact on the coastal provinces. Of course, 100 per cent import interdiction cannot be achieved overnight and may never be achieved at all, given the willingness and capability of neighbouring countries to revert to rail imports, however marginal. Interdiction of 90 per cent of oil imports is not only achievable, but impossible to offset through other transport means.

This will force the People's Republic of China to rely on its strategic reserve almost immediately and cause a massive reallocation of fuel use requirements. It may also have localized impacts on military forces, as it will be much harder to supply PLAN and PLAAF units based in the south. Only two of the Strategic Petroleum Reserve depots are in the south, comprising less than 20 percent of the Strategic Petroleum Reserve.

Additional effects on internal energy transport are another element of the strategy. The inshore effort is intended to disrupt both military and energy logistics. In the case of coal, 30 percent of domestic coal transport is by river and coastal traffic, which is especially vulnerable to mine warfare. Chinese short-haul shipping is a commercial and not a state enterprise, and civilian shipowners have been traditionally unwilling to risk their vessels in hostile waters. A ship sunk at a loading berth blocks the facility effectively and for a significant duration.

Infrastructure degradation will affect both water and rail transport, especially if rail bridges are dropped into major waterways. The Danube River was effectively closed to large traffic for five years after the Novi Sad bridges were dropped in Operation Allied Force. Damage to pipeline pumping stations, rail tunnels, bridges and refineries will be time consuming and difficult to repair, and in the case of refineries, suitable equipment may not be available domestically.

The secondary effects on electricity production will likewise ripple through the transportation and industrial sectors.

Electricity shortages caused by oil or coal interdiction will affect the train network; refineries starved of either feedstock or electricity cannot refine and pipelines without electricity do not move oil. Reduced diesel production will affect the non-electric portion of the rail network plus both maritime and truck transport, while at the same time diesel will be in demand for emergency power generation.

Reprioritization of limited freight transport will affect industry (itself starved for power) and agriculture directly, as well as disrupting distribution of industrial or agricultural products. Local surpluses and shortages of fuel, coal and electricity are certain to occur, further complicating distribution challenges.

Similar effects can be directly observed from single industrial accident. In November of 2013 a Sinopec pipeline in Huangdao, Shandong Province exploded, killing over 60 people and shutting the pipeline down. This caused production cutbacks in two nearby refineries, a reallocation of refinery production company-wide, and a shutdown of the Qingdao oil terminal for a week. Tankers were diverted to other ports, causing offshore backups because of the lack of available offload facilities. Environmental damage took many weeks to clean up and the oil berths were out of commission for months.

All of these cascading events were the result of the equivalent of a single weapon hit and the pipeline was never repaired.

The duration of any campaign is difficult to predict. The amount of military storage for refined fuel remains an unknown factor. Similarly, there are absolute limits on refinery production, rail transport, and truck movement of refined products, none of which are known, perhaps even to the People's Republic of China government. Finally, the wartime consumption of jet fuel by the PLAN and PLAAF is largely conjectural. Further complicating any assessment is the fact that turbine-powered ships can and do run on marine diesel fuel, which is still refined distillate, but is closer to diesel fuel in composition than kerosene.

A counter-logistics campaign, fought from long range where possible, is intended to provide a strategy that avoids China's strengths in air defense and relies on a very limited target list focused on targets that are neither hardened nor mobile.

Instead of matching technologically advanced military forces against like systems in terrain favourable to China, it is intended to fight only those units that come out to fight and leave many of their advantages behind.

This is a deliberate offset strategy, tailored to China, which avoids the pitfalls inherent in the misapplication of older air power theory and takes the specific characteristics of the adversary into account. It is also a strategy that could be executed today, with today's force structure, posture and today's personnel.

The Pentagon could certainly improve in all of those areas, but the execution of a Strategic Interdiction campaign will not need to wait for the development of new technologies and it does not hinge on transient vulnerabilities.

Our experience in World War II demonstrated the effectiveness of our efforts to successfully interdict the Japanese transportation systems and oil storage and production facilities. The Pacific Strategic Bombing Survey noted in retrospect that our efforts were inefficiently directed — if we had possessed accurate intelligence about the nature of Japan's logistics network, we might have rearranged our targeting priorities to increase our effects and shortened our timelines.

With respect to China, we do have significant knowledge about the energy sector, precisely because it is involved directly in foreign trade and a great deal of data is available. Instead of attempting to fight a generic "near peer" adversary with a template drawn from Desert Storm, we should be planning to apply a counter-logistics strategy against a real adversary, with the attendant national characteristics, vulnerabilities and geography.

Colonel Michael Pietrucha, USAFR (written as a duty assignment for the US Government, this essay does not reflect Government policy)

NOTES

INTRODUCTION:

1. Samuel Huntington, *The Clash of Civilisations and the Remaking of the World Order* (Simon and Schuster, 1996)

CHAPTER ONE: Australian Army

1. Pugach, I. et al 2013, *Genome-wide data substantial Holocene gene flow from India to Australia*, Proceedings of the National Academy of Sciences, vol. 110, no. 5, 1803-1808
2. Australian National Audit Office, *Multi-Role Helicopter Program*, Audit Report No.52 2013-14

CHAPTER TWO: Royal Australian Navy

1. Naval Graphics, *Submarines of the World*

CHAPTER THREE: Royal Australian Air Force

1. Colin Clark, *Gen. Mike Hostage on the F-35: No Growlers Needed When War Starts*, Breaking Defense, 6th June, 2014
2. Center For Defense Information at the Project on Government Over-sight, *Not Ready for Prime Time*
3. David Axe, *Joint Strike Fighter's Outrageous Claim*, Wired 22nd September, 2008

CHAPTER FOUR: Fuel Security

1. Henrik Svensmark and Nigel Calder, *The Chilling Stars* (Icon Books, 2007)

CHAPTER FIVE: China's Coming War

1. Francis Fukuyama, *The End of History and the Last Man* (Free Press, 1992)

2. Samuel Huntington, *The Clash of Civilisations and the Remaking of the World Order* (Simon and Schuster, 1996)
3. Edward Lutwark, *The Rise of China vs. the Logic of Strategy* (Belknap Press, 2012)
4. John Lee, 31st December, 2014, *What Surging Chinese Investment in Australia Says about China*, Hudson Institute

CHAPTER SIX: Going Nuclear

1. Amir Taheri, *The Persian Night* (Encounter Books, 2009)
2. Paul Bracken, *The Second Nuclear Age* (Times Books, 2012)

CHAPTER SEVEN: The Broader Strategic Context

1. Victor David Hansen, *Carnage and Culture* (Anchor, 2002)
2. Thomas Barnett, *The Pentagon's New Map* (Berkley Trade, 2005)
3. Niall Ferguson, Civilisation: *The West and the Rest* (Penguin Books, 2012)
4. Jeffrey A. Tucker, "When Capital is Nowhere in View," Mises Daily, Ludwig von Mises Institute, May 10, 2011, http://mises.org/daily/5277/

CHAPTER EIGHT: Funding the Increased Defence Effort

1. Steering Committee for the Review of Government Service Provision, *2014 Indigenous Expenditure Report*
2. First Principles Review, 2015, *Creating One Defence*